Issues in Bioinvasion Science

The editors wish to acknowledge the following referees:

Benigno Elvira, *Universidad Complutense de Madrid, Spain*
José Esteban Durán, *Instituto Nacional de Investigación y Tecnología Agraria y Alimentaria, Spain*
Piero Genovesi, *National Wildlife Institute, Italy*
Félix Llamas García, *Universidad de León, Spain*
Marco Masseti, *Università degli Studi di Firenze, Italy*
Nicolás Pérez Hidalgo, *Universidad de León, Spain*
María Antonia Ribera Siguan, *Universidad de Barcelona, Spain*
Daniel Simberloff, *University of Tennessee, USA*
Daniel Sol, *McGill University, Canada*

Issues in Bioinvasion Science

EEI 2003: a Contribution to the Knowledge on Invasive Alien Species

Edited by

Laura Capdevila-Argüelles and Bernardo Zilletti
G.E.I. – Grupo Especies Invasoras, C.B.I. – Centro Invasiones Biológicas, León, Spain

Reprinted from *Biological Invasions,* Volume 7(1), 2005.

Springer

A C.I.P. catalogue record for this book is available from the Library of Congress

ISBN 978-90-481-6745-6

e-ISBN 978-1-4020-3870-9

Published by Springer,
P.O. Box 17, 3300 AA Dordrecht, The Netherlands

Sold and distributed in North, Central and South America
by Springer,
101 Philip Drive, Norwell, MA 02061, USA

In all other countries, sold and distributed
by Springer,
P.O. Box 322, 3300 AH Dordrecht, The Netherlands

Cover photo: Mustela vison (American mink), *Esox lucius* (northern pike), *Trachemys scripta* (red-eared slider), *Myocastor coypus* (nutria), *Estrilda astrild* (common waxbill) and *Poecilia reticulata* (guppy). Photocomposition by G.E.I.

Printed on acid-free paper

Printed in the Netherlands

Table of contents

Introduction

The prevention and minimization of the impact of invasive alien species has become a priority on the agenda of the international community because of their negative impact on biodiversity, economy and public health. This has led in recent years to the development of tools (legal, technical, etc.) and programmes. It has become clear that the scale of the problem requires an interdisciplinary approach, the planning of long-term actions (strategic perspective), and urgency.

Biological invasions are a global problem whose local impact can be of great magnitude. Effective responses are needed, cooperation and coordination among countries and/or different sectors of society are a must. In this frame, scientific research is fundamental to improving knowledge on invasive alien species and developing new methodologies to fight them.

Big steps to improve the understanding of biological invasions have been taken in recent years thanks to the increase of basic research in this field. However, in spite of the successes obtained and enlightening information gained on 'the biological invasions phenomenon', there are still many gaps to be filled. The implementation of models, the improvement of mitigation methods (more effective and more environmentally friendly) and prevention techniques (e.g. risk analysis) are a priority. However given that invasive alien species represent a global threat, it is necessary to make the best use of the current information available.

It is absolutely necessary to optimise the efforts in scientific research, establishing mechanisms to develop the exchange of information on invasive alien species, keeping in mind that in the case of biological invasions solutions to a problem may be found in other parts of the world.

In this spirit and with the objective, among others, of combining and sharing the experiences on biological invasions carried out in Spain, the GEI (Grupo Especies Invasoras) organised the First National Conference on Invasive Alien Species in 2003. More than 150 people from different parts of the peninsula and from other countries belonging to different institutions (universities, public administration, NGOs) responded to this 'call for action' and participated in the conference.

The present publication, which contains some of the contributions of the First National Conference on Invasive Alien species, aims to increase global knowledge on invasive alien species.

For editing this issue we counted on the help of a large number of people. We are grateful to the authors for their contributions and full response to the suggestions and requirement of the editors. The committee of referees played a very important role in the project. To all of them, for their patient work, invaluable intellectual contribution and support, our most sincere thanks. Our gratitude goes to James T. Carlton and James Drake, who have succeeded each other in the scientific direction of *Biological Invasions*, for having granted us the opportunity to publish the present issue and for supervising its scientific quality. We owe thanks to Springer and, in particular, to Suzanne Mekking and Ellen Girmscheid who have helped bring this project to completion.

We are deeply indebted to all the institutions who sponsored the conference, especially to Fundación Biodiversidad, which funded, among other things, the present edition.

Thanks are also due to the GISP (Global Invasive Species Programme), ISSG/IUCN (Invasive Species Specialists Group/World Conservation Union) and COE (Council of Europe), who have supported our work and participated in the conference.

Last, but not least, we extend our warmest thanks to Maj de Poorter, Piero Genovesi, Michel Pascal, Steve Raaymakers, Eladio Fernández-Galiano Daniel Simberloff, Richard N. Mack and Ana Isabel Queiroz for their constant and ever present help and support.

Laura Capdevila-Argüelles & Bernardo Zilletti
G.E.I. – Grupo Especies Invasoras
c/ Moisés de Léon N. 22 Bajo
24006 Léon
Spain
E-mail: geiinvasoras@usuarios.retecal.es

Funded by

Fundación Biodiversidad

MINISTERIO DE MEDIO AMBIENTE
SECRETARIA GENERAL DE MEDIO AMBIENTE
DIRECCION GENERAL DE CONSERVACION DE LA NATURALEZA

Caja España
OBRA SOCIAL
Alta rentabilidad social

Other sponsors

Port de Barcelona
Autoritat Portuària de Barcelona

Junta de Castilla y León
CONSEJERIA DE EDUCACION Y CULTURA

UNIVERSIDAD DE LEÓN

Scientific patronage

COUNCIL OF EUROPE CONSEIL DE L'EUROPE
Bajo el patrocinio del Sr. Walter SCHWIMMER, Secretario General del Consejo de Europa

GISP
Global Invasive Species Programme

ISSG
Invasive Species Specialist Group

IUCN
The World Conservation Union

Biological Invasions (2005) 7: 3–15

Invasion biology of Australian ectomycorrhizal fungi introduced with eucalypt plantations into the Iberian Peninsula

Jesús Díez
Departamento Biología Vegetal, Universidad de Alcalá, 28871 Alcalá de Henares (Madrid), Spain (e-mail: diez_muriel@yahoo.com; fax: +34-91-8855066)

Received 4 June 2003; accepted in revised form 30 March 2004

Key words: ectomycorrhizal fungi, *Eucalyptus*, exotic fungi, forest plantations, invasion ecology

Abstract

In the last two centuries, several species of Australian eucalypts (e.g. *Eucalyptus camaldulensis* and *E. globulus*) were introduced into the Iberian Peninsula for the production of paper pulp. The effects of the introduction of exotic root-symbitotic fungi together with the eucalypts have received little attention. During the past years, we have investigated the biology of ectomycorrhizal fungi in eucalypt plantations in the Iberian Peninsula. In the plantations studied, we found fruit bodies of several Australian ectomycorrhizal fungi and identified their ectomycorrhizas with DNA molecular markers. The most frequent species were *Hydnangium carneum*, *Hymenogaster albus*, *Hysterangium inflatum*, *Labyrinthomyces donkii*, *Laccaria fraterna*, *Pisolithus albus*, *P. microcarpus*, *Rhulandiella berolinensis*, *Setchelliogaster rheophyllus*, and *Tricholoma eucalypticum*. These fungi were likely brought from Australia together with the eucalypts, and they seem to have facilitated the establishment of eucalypt plantations and their naturalization. The dispersion of Australian fungal propagules may be facilitating the spread of eucalypts along watercourses in semiarid regions increasing the water lost. Because ectomycorrhizal fungi are obligate symbionts, their capacity to persist after eradication of eucalypt stands, and/or to extend beyond forest plantations, would rely on the possibility to find compatible native host trees, and to outcompete the native ectomycorrhizal fungi. Here we illustrate the case of the Australasian species *Laccaria fraterna*, which fruits in Mediterranean shrublands of ectomycorrhizal species of *Cistus* (rockroses). We need to know which other Australasian fungi extend to the native ecosystems, if we are to predict environmental risks associated with the introduction of Australasian ectomycorrhizal fungi into the Iberian Peninsula.

Introduction

Alien plants often require mutualistic partners to overcome barriers to establishment in foreign environments. Mutualisms that facilitate invasions occur at several phases of the life cycle of alien invading plants. However, even when wind and native generalist animals mediate flower pollination and seed spread, the lack of compatible mycorrhizal symbionts can limit the spread of alien plants (Richardson et al. 2000a).

Mycorrhizal fungi are essential in plant nutrition in terrestrial ecosystems, and terrestrial plants present different types of obligate mycorrhizal symbioses. Herbs and shrubs form arbuscular mycorrhizas (AM) with glomales (Smith and Read 1997). The low specificity of AM relationships and the easy acquisition of mutualistic symbionts by herbs and shrubs in any ecosystems are important reasons for so many ecosystems being susceptible to invasion by alien plants. As a consequence, the introduction of AM fungi does not seem to play a major role in mediating plant invasions, except on some remote islands that are poor in AM fungi (Richardson et al. 2000a). Most forest trees, however, associate with a group of basidiomycetes and ascomycetes forming ectomycorrhizas (ECM) (Newman and Reddell 1987). Ectomycorrhizal symbioses present a range of host–fungus specificities. For this reason, exotic forestry has often needed the introduction of compatible ectomycorrhizal fungal symbionts (Grove and Le Tacon 1993). For many non-native trees, notably for pines and eucalypts, the lack of symbionts was a major barrier to establishment and invasion in the southern hemisphere, before the build-up of inoculums through human activity (Armstrong and Hensbergen 1996; Davis et al. 1996; Richardson et al. 2000a). The role of ECM fungi in facilitating the establishment and invasion of alien trees was claimed by Richardson et al. (1994) to explain the invasion patterns of pines in South Africa. Little investigation, however, has been done on the role played by the ECM fungi in the naturalization of eucalypts beyond their natural range.

This paper provides a framework for thinking about the effects of the introduction of exotic fungi with plantations of exotic forest trees. Different sections will deal with aspects of the invasion biology of eucalypts and their ECM fungi. We first lay the groundwork for our paper by briefly introducing the history of the exotic forestry and several terms used in studies of biological invasions. The paper will be illustrated with a study conducted on exotic plantations of eucalypts in the Iberian Peninsula. We will describe ectomycorrhizal communities of Australasian fungi present in these plantations. We will next analyze the role that ECM fungi plays in promoting (or limiting) invasion rates of the eucalypts introduced. The central core of the paper will try to explain the lags between the introduction and the spread of the eucalypts, as a factor that depends on the dispersion of propagules of introduced ECM fungi. We report host shifts of Australian fungi to native ectomycorrhizal plants detected to date. We discuss whether the introduced fungi could threaten natural communities of ECM fungi by out-competing the native ECM fungi from their natural hosts. This work includes an analysis of the potential effects of the invasion of these exotic fungi in the nutrient cycling of Mediterranean forests. To the best of our knowledge, this is the first investigation dealing with the invasion ecology of exotic ECM fungi.

Exotic forestry: invasions of alien trees

Origin of exotic forestry

There is evidence of large-scale forestation in the ancient Mediterranean basin, where timber- and crop-producing trees were planted as long ago as 255 B.C. (Zobel et al. 1987). Inspite of its long history, the scale of forestry remained small until recently. Large-scale forestry was not widespread until the second half of the 20th century (Zobel et al. 1987), when pines and eucalypts were widely planted outside their natural ranges. Pines, which comprise only Holarctic species, were planted in South America, South Africa and Australia. The Australasian eucalypts were planted worldwide. In particular, the need for increased wood production to improve living conditions made the genus *Eucalyptus* one of the most widely planted silvicultural crops. In addition, in many damaged ecosystems, afforestation with alien eucalypts was driven by the belief that such plantings were beneficial to the environment (Zobel et al. 1987).

Naturalization and invasion of pines and eucalypts

In this paper, we will use the following three concepts: introduction, naturalization and invasion, as defined by Richardson et al. (2000b). A tree introduction takes place when humans transport a tree across a geographical barrier to a new area. Naturalization refers to the species establishing new self-perpetuating populations and becoming incorporated within the native flora. The naturalized trees regenerate freely, but mainly under their own canopies. In contrast, invasive species recruit seedlings, often in very large numbers, at long distances from parent plants (often more than 100 m). Only some of the naturalized plants become invasive, producing important environmental or economical damages.

All trees that are widely planted in alien environments can become invasive and spread under certain conditions (Richardson 1998). Consequently, the use of exotic trees has often caused environmental damages in different parts of the world (Binggeli 1996). The species that cause the greatest problems are generally those planted most widely and for the longest time (Pryor 1991; Rejmánek and Richardson 1996). According to Higgins and Richardson (1998), at least 19 *Pinus* species are invaders of natural ecosystems in the southern hemisphere; four of the most widespread invasive pines are *P. halepensis*, *P. patula*, *P. pinaster* and *P. radiata* (Higgins and Richardson 1998; Richardson et al. 1990).

The genus *Eucalyptus* L'Hérit (Myrtaceae) comprises evergreen woody plants, including shrubs and forest trees (nearly 600 species), which are confined in natural occurrence entirely to the Australasian region, Papua, New Guinea and Timor (Pryor 1976). More than 43 species of eucalypt trees are planted outside their natural geographic distribution. Eucalypts are planted on a large scale to provide a short rotation crop yielding wood and paper pulp for industrial use (Eldridge et al. 1994). Although less invasive than pines, several species of eucalypts already caused problems as invaders in South Africa (Richardson 1998; Richardson et al. 2000a). Eucalypts are represented on many weed lists from other parts of the world, including California (Warner 1999) and peninsular Spain (Sanz-Elorza et al. 2001). Significant impacts might result from the introduction of eucalypts into Spain and some transformation on various ecosystem properties; specially changes in grasslands and scrubland habitats. Ecology and environmental politics dictate the desirability of maintaining Mediterranean grasslands and scrublands due to their high diversity in endemic plants and because these ecosystems are the natural habitats for many local wildlife. Eucalypt invasions can cause shifts in life-form dominance, reduced diversity, disruption of prevailing vegetation dynamics, and changing nutrient cycling patterns. Hence, eucalypt plantations are increasingly causing major conflicts between Spanish foresters, politicians and conservationists.

Role of ectomycorrhizal fungi in exotic forestry

Mycorrhizas of herbs, shrubs and forest trees

Mutualistic interactions between fungi and plant roots are common in the plant kingdom, including mycorrhizal symbioses. In most mycorrhizal symbioses, the fungal partner supplies nutrients to the host plant in exchange for photosynthetic carbon, and may offer protection against pathogens, toxins and drought (Smith and Read 1997). The main types of mycorrhizal symbioses differ in the anatomy of the mycorrhiza, which is a mixed root-fungus structure at which nutrient interchanges take place (Smith and Read 1997). Most Mediterranean plants are mycorrhizal with different types of fungi. Herbs and shrubs mainly associate with glomales to form arbuscular mycorrhizas (AM), which is by far the most common mycorrhizal symbiosis in the Mediterranean terrestrial ecosystems (Díez 1998). A restricted group of plants form particular types of mycorrhizas with a range of particular ascomycetes and basidiomycetes, such as the Ericales and Orchidales. However, most forest trees form ectomycorrhizas (EM) with a polyphyletic group of basidiomycetes and ascomycetes (Smith and Read 1997; Hibbett et al. 2000).

Forest trees are obligate ectomycorrhizal plants

Pines and eucalypts are obligate ectomycorrhizal trees, which depend on these mutualistic symbioses for nutrient uptake in natural conditions (Smith and Read 1997). Ectomycorrhizal symbioses are essential in the mobilization of nutrients in soil forests (Fahey 1992; Read and Pérez-Moreno 2003). In natural conditions, these forest trees rely on the ectomycorrhizal fungi, which colonize the root cortex and form a nutrient-gathering 'organ' called ectomycorrhiza (Smith and Read 1997). Ectomycorrhizas are almost exclusive to forest trees. In the Mediterranean, only a restricted group of shrubs in the Cistaceae (*Cistus* spp., called rockroses) forms EM, in which are involved a range of endemic ascomycetes and basidiomycetes (Díez 1998). Ectomycorrhizas are characterized structurally by the presence of a dense mass of fungal mycelium surrounding the short lateral roots (the mantle). The mantle originates from the attachment of the fungal hyphae onto epidermal cells, and the multiplication of hyphae to form a series of hyphal layers. The fungal mycelium also grows among the cortical cells, forming the Hartig net. The Hartig net is the structural and functional interface between fungal and roots cells. The fungal mantle is connected with a highly extended network of mycelium prospecting the soil and gathering nutrients. The extra-radical mycelium is responsible for mobilizing soil nutrients, nutrient (and water) uptake and transfer to the ectomycorrhiza (Peterson and Bonfante 1994). Sclerotia (and sclerotia-like bodies) are vegetative balls of hyphae formed by a few species of ECM fungi. Finally, fruiting bodies arise from discrete points of the extra-radical mycelium to ensure the sexual reproduction and the dispersal of the fungal partner (Allen 1991).

The seedlings of forest trees need to be colonized by ectomycorrhizal fungi. In a given biotope, the seedling survival depends on the presence of the propagules of compatible ectomycorrhizal fungi in the soil (spores, sclerotia, or soil mycelium). Hyphae arising from fungal propagules have a limited capacity to grow and die unless they come in contact with a root tip of a compatible host. In the forest, seedlings grow near mycorrhizal trees and may thus become colonized by pre-existing mycorrhizal mycelia of the living roots of mature trees (Onguene and Kuyper 2002). In open areas, the fungal hyphae and then the sclerotia disappear in the long term in the absence of compatible ECM plants (Brundrett and Abbott 1995). Thus, the colonization of the root seedlings in a new biotope necessarily depends on the spore dispersion from close forests (Allen 1991). Spores of epigeous (fruiting aboveground) fungi are mainly dispersed by wind, whereas mycophagous animals are important vectors for dispersing spores of hypogeous (fruiting below ground) fungi.

Success of exotic forest plantations: the occurrence of compatible ectomycorrhizal fungi

In the AM symbiosis, the levels of specificity among host plants and fungal species are low, and many glomales have a cosmopolitan distribution (Smith and Read 1997). Due to this low specificity, most invading herbs and shrubs have no problems to form mycorrhizas with the fungi of the target habitat (Richardson et al. 2000a). In contrast, the ECM fungi present different levels of specificity. A given species of ectomycorrhizal fungus is usually only able to establish mutualistic symbiosis with a number of species from the same biogeographic realm, and even more highly specific interactions occur (Molina et al. 1992). Specially, Australian ectomycorrhizal plants (e.g. *Eucalyptus*, *Acacias* spp.) have evolved in isolation from the ECM fungal flora associated with *Pinus* (and *Quercus)* in the northern hemisphere (Halling 2001). This could explain why many (if not almost all) native fungi from the northern hemisphere do not associate with eucalypts *in silva*, and *vice versa*.

When a compatible ectomycorrhizal biota is absent on the plantation site, there is a barrier for the success of exotic plantations. For this reason, mycorrhizal inoculation of pine and eucalypt seedlings with forest soil, spores or mycelium of compatible ECM fungi was often necessary for the success of exotic plantations of pines (Perry et al. 1987; Grove and Le Tacon 1993). In the Southern Hemisphere, pine forestry was delayed by the lack of suitable ECM fungi until a number of Holarctic ECM fungi were introduced with the pines (Dunstan et al. 1998). Some of these exotic pines eventually invaded a wide range of systems with the introduction of such Holartic ECM fungi (Richardson et al. 2000a).

Study areas and methods used to identify the ectomycorrhizal fungi

Plantations of eucalypts in Spain

Since the last century, the river red gum *Eucalyptus camaldulensis* Dehnh. and the blue gum *E. globulus* Labill. have been used in extensive plantations in the Iberian Peninsula. Nowadays, there are around 550,000 ha of eucalypts plantations in Spain, 320,000 ha of *E. globulus* and about 180,000 ha of *E. camaldulensis*. *Eucalyptus camaldulensis* is planted mainly in southwestern Spain (Huelva, Cadiz, Badajoz and Seville), and *E. globulus* in the northern regions (Galicia, Asturias and Santander). *Eucalyptus gomphocephala* DC is used on basic soils in Murcia and Almería (southeastern Spain). These eucalypts are used also in afforestation and agroforestry (windbreaks, shelter trees, and intercropping of trees and arable crops). In the Iberian Peninsula, the inoculation with ectomycorrhizal fungi was not necessary to ensure the success of many eucalypt plantations. Two hypotheses might account for the lack of need of inoculation: (i) whether there were native ectomycorrhizal fungal species compatible with the eucalypts, (ii) or a range of ectomycorrhizal fungi native from Australasia were brought together with the eucalypt seedlings. To resolve this question, we have been investigating the origin of ectomycorrhizal fungi present in the Iberian plantations of eucalypts.

Study area and sampling strategy

The study area is located in the region of Extremadura, which is formed by the provinces of Cáceres and Badajoz, where extensive plantations with eucalypts took place between 1955 and 1977. *Eucalyptus camaldulensis* was by far the most predominant species, followed by *E. globulus*. Extremadura stands out with 14% of the totality of Spanish plantations of eucalypts. We studied eucalypts stands, shelterbelts and road verges. We sampled watercourses and riparian stands in which eucalypts became naturalized and invasive.

We sampled fruit bodies and ectomycorrhizal root tips. The fruit bodies are identified using morphological features. In some cases, the use of molecular and phylogenetic methods is necessary to discriminate among cryptic species (Díez et al. 2001). We collected ectomycorrhizal roots, because many ectomycorrhizal species do not fruit, and fruiting patterns do not truly reflect the belowground community of ECM fungi (Gardes and Bruns 1996; Horton and Bruns 2001).

Ectomycorrhiza identification: morphological and molecular methods

Despite the general organization, ectomycorrhizas differ among species in colour, mycelium density, size, forms and biochemical composition. Morphological typing of ectomycorrhizas enables us to identify fungi that seldom or never produce fruiting bodies. Methods for the morphological characterization of ectomycorrhizas are described in Ingleby et al. (1990) and Agerer (1997), who provided descriptions of ectomycorrhizas and criteria to discriminate species based on morphological features and chemical tests. Typing ectomycorrhizas with morphological methods is time consuming; and in many cases, it is not conclusive, because the ectomycorrhizas have not been described for many ECM fungi. In addition, the ectomycorrhizas of one fungal species can also show different morphologies according to the host, physiological and environmental conditions (Egger 1995). With such a morphological approach, we can classify the ectomycorrhizal tips as much as in morphotypes.

To overcome the problems of morphological typing, we use molecular methods for the identification of ectomycorrhizas, as described in Martín et al. (2000). Such methods are based on the restriction fragment length polymorphism (RFLP) of the internal transcribed sequences (ITS) of the nuclear rDNA. The ITS regions is amplified for ectomycorrhizal root tips with the polymerase reaction technique (PCR), using a thermostable DNA polymerase and primer pairs annealing at conserved regions of the 18S and 28S ribosomal genes (White et al. 1990). Such PCR-based methods are of great value in these kinds of studies (Glen et al. 2001a, b). We amplify the ITS regions from DNA obtained from ECM root tips, using fungal-specific primers to avoid the amplification of plant DNA (Gardes et al. 1991; Gardes and Bruns 1993). After cutting the PCR-amplified ITS with restriction enzymes, we obtain RFLP patterns.

To identify the different ectomycorrhizas, we are compiling a database of RFLP profiles of fruit bodies, so that we can compare them with those obtained from ECM roots. This PCR-RFLP database is of great help in the identification of unknown ectomycorrhizas. Most fungal species show a unique RFLP pattern, and their ectomycorrhizas can be identified by their PCR-RFLP profiles. For RFLP patterns not associated with any of the fruit bodies, direct sequencing of the ITS regions followed by a 'Blast research' (Altschul et al. 1997) in the National Centre for Biotechnology Information (http://www2.ncbi.nlm.nih.gov/) enable us to identify the ectomycorrhizal morphotypes, with luck, even at the species level.

Ectomycorrhizal fungi of eucalypt plantations in the Iberian Peninsula

Species of introduced Australian ectomycorrhizal fungi

Over our surveys in Iberian plantations of eucalypts, we found fungi known only from Australian forests and eucalypts plantations worldwide

Table 1. Twelve frequent Australasian fungi that have been introduced in the Iberian Peninsula together with the eucalypts.

Taxonomic group	Species	Habit
Basidiomycetes	*Laccaria fraterna*[a]	Epigeous
	Hydnangium carneum[a]	Hypogeous
	Hymenogaster albus[a]	Hypogeous
	Hysterangium inflatum[a]	Hypogeous
	Pisolithus albus[a]	Epigeous
	Pisolithus microcarpus[a]	Epigeous
	Setchelliogaster rheophyllus[a]	Semi-hypogeous
	Tricholoma eucalypticum[a]	Epigeous
Ascomycetes	*Labyrinthomyces donkii*[a]	Hypogeous
	Rhulandiella berolinensis[a]	Hypogeous
	Discinella terrestris[c]	Epigeous
	Urnula rhytidia[b]	Epigeous

The table also includes a saprophytic fungus and a possible facultatively ectomycorrhizal fungus of Australasian origin.
[a] Ectomycorrhyzal.
[b] Saprophytic, and probably facultatively ectomycorrhizal.
[c] Saprophytic.

(Table 1). Our analyses of ECM roots confirmed the presence of Australian ECM fungi in the eucalypt roots of Spanish plantations. Surveys of fruit body proved that fungi fruiting in the Iberian plantation of eucalypts are of Australian origin.

The most frequently occurring species in the Spanish plantations are *Hydnangium carneum* Wallr., *Hymenosgaster albus* (Klotzsch) Berkeley and Broome, *Hysterangium inflatum* Rod. (= *H. pterosporum* Donadini and Riousset), *Labyrinthomyces donkii* Malen ç. *Pisolithus albus* (Cke and Mass.) M.J. Priest, *P. microcarpus* (Cke and Mass.) Cunn., *Ruhlandiella berolinensis* (Henn.) Diss. and Korf, *Laccaria fraterna* (Cooke and Mass.) Sacc. (*L. lateritia* Malenç), *Setchelliogaster rheophyllus* (Ber. and Malenç.) G. Moreno and Kreisel, and *Tricholoma eucalypticum* Pearson. Most of these Australian fungi are able to fruit under the climatic conditions of the Iberian Peninsula. Additional ECM fungi, likely unable to fruit outside their natural range, were detected belowground during the molecular typing of ectomycorrhizal roots. We identified ectomycorrhizas of *Cenococcum geophilum* Fr. and several species of *Sebacina* and *Thelephora*. Whether the *C. geophilum*, *Sebacinia* and *Thelephora* strains infecting eucalypts in the Iberian plantations are natives or Australian deserves further studies, because these species are known to be native to the area studied.

In Extremadura, the soil dries up in summer, and most fungi in the eucalypt plantations fruit in late winter and spring. Some of these species are secotioids or truffle-like (sequestrate) fungi, most of them adapted to fruit belowground (hypogeous fungi). *Hydnangium carneum* is one of the most common truffle-like fungi in the eucalypt stands; which is a gastroid relative of the agaricoid genus *Laccaria;* due to its hypogeous habit, this Australian fungus is well adapted to the conditions of the Mediterranean climate. We also found the Australian false truffle *Hymenogaster albus,* which is common in plantations near the Monfragüe Natural Park. *Hysterangium inflatum* is hypogeous relative to *Phallus*, very common in the Monfragüe region. *Setchelliogaster rheophyllus* is another secotioid fungus present in eucalypt plantations near Monfragüe and Badajoz. *Rhulandiella berolinensis* is considered specific to Australasian ectomycorrhizal plants, and is very common in riparian stands of *Eucalyptus camaldulensis* near Mérida (Badajoz). The main epigeous fungus (fruiting aboveground) was *Laccaria fraterna*, which is an agaricoid species native to Australia and introduced into the Mediterranean with eucalypts. To the best of our knowledge, our collections of *Tricholoma eucalypticum* are among the first records of this fungus outside Australia.

Growing on debris of eucalypts and humus of eucalypt plantations, we found the Australasian species *Urnula rhytidia* (Berk.) Cooke. (Pezizales) and *Discinella terrestris* (Berk. and Br.) Dennis (Leotiales); these two species are regarded as characteristic of sclerophyllous eucalypt forests of Australia and Tasmania. There is no reliable information on whether these fungi are saprophytic or facultatively ectomycorrhizal as many other pezizales.

In the studied plantations of eucalypts in Extremadura, we found strains of the Australasian species *Pisolithus albus* and *P. microcarpus*, as proved with molecular analyses of the ITS sequences of the nuclear rDNA (Díez et al. 2001). For many years, the name *Pisolithus arrhizus* (Pers.) Rauscher (synonym of *P. tinctorius* [Pers.] Coker and Couch) have been used for all *Pisolithus* strains occurring in eucalypt and pine plantations worldwide, regardless of the host plant (Cairney 2002). This misunderstanding occurred because many researchers considered

Pisolithus as a monospecific genus, and *P. arhizus* as a fungus with a wide host range (Chambers and Cairney 1999). However, our molecular analyses proved that the genus *Pisolithus* comprises several phylogenetic species, and that each species of *Pisolithus* is confined to hosts from one single biogeographic realm. Our study also proved that *Pisolithus arhizus* is restricted to Holarctic host plants (e.g. *Quercus* and *Pinus* spp.) and does not occur in eucalypt plantations (Martin et al. 2002). In a previous work (Díez et al. 2001), we showed that *Pisolithus albus* and *P. microcarpus* fruit in litter and on open ground and at the edges of eucalyptus plantations, and on dry and disturbed sites such as gravelly roadsides in Morocco and Spain. Endemic to Australia, *P. albus* and *P. microparpus* are not restricted to eucalypts and form ectomycorrhizas with other Australasian plants; we have found these two species in association with Australasian acacias in Portugal (Muriel and Díez, unpublished). The species of *Pisolithus* native to the Iberian Peninsula correspond to *P. tinctorius*, and two unnamed species, one basophilic species and another *Cistus*-specific *Pisolithus* species (Díez et al. 2001); these three Holarctic species of *Pisolithus* never occur in association with eucalypts (Díez et al. 2001; Martin et al. 2002).

A group of Australasian ectomycorrhizal fungi were introduced with the eucalypts

The fungi we found in the Iberian plantations of eucalypts are of Australian origin. These exotic fungi were likely introduced with eucalypt seedlings brought into peninsular Spain before plant quarantine restrictions were observed. In Australia, gum seedlings are container grown in nurseries, which are naturally colonized by a limited number of ectomycorrhizal fungi. Most of these ECM fungal species may persist during the first years of eucalypt plantations (Lu et al. 1999). Foresters probably dispersed these exotic ECM fungi in soil or eucalypt seedlings worldwide. Our results are in agreement with investigations by other mycologists, and there is a growing consensus in a worldwide dispersal of a number of Australasian ECM fungi together with the eucalypts (Giachini et al. 2000). Saprophytic and even pathogenic fungi seem to have spread worldwide as well with the eucalypts. In this regard, it has been suggested that dissemination of the basidiomycetous yeast *Cryptococcus neoformans*, a human pathogen associated with *Eucalyptus* leaves in southern California and India resulted from the introduction of eucalypts (Casadevall and Perfect 1998; Chakrabarti et al. 1997).

Do native fungi infect exotic eucalypts in sylva?

In the Iberian Peninsula, there are many native ECM fungi in association with pines, oaks, and a restricted number of ectomycorrhizal shrubs (i.e. *Cistus* spp.) (Díez 1998). In our surveys, we did not find Holarctic ECM fungi in the eucalypt plantations, though propagules of native fungi are often present on planting sites. We do not know any reliable evidence of European fungi forming ectomycorrhizas with eucalypts in natural conditions in Spain. Because eucalypts have evolved in isolation from the ectomycorrhizal mycobiota associated with *Pinus* and *Quercus* in the Holarctic Realm, the native fungi (e.g. *Pisolithus tinctorius*) might be unable to associate with eucalypts *in silva*. Even within each biogeographic realm, there would exist some level of host-symbiont specificity. Parladé et al. (1996) described, in pure culture syntheses, the ability of native Iberian fungi to colonize several North American conifers planted in northern Spain. Such an ability can be easily explained by the similarities between the fungal floras of North America and the Iberian Peninsula (Halling 2001), as these two regions belong to the Holarctic Realm. However, some host specificity might exist, because in tree nurseries and experimental plantations in northern Spain, native ECM fungi do not seem to outcompete exotic North American fungi inoculated (or that accidentally infect seedlings) in tree nurseries (Pera et al. 1999).

Role of the introduction of Australasian ectomycorrhizal fungi in promoting eucalypt invasiveness

We now have evidence on the introduction of Australian ectomycorrhizal fungi with the plantations of eucalypts in the Iberian Peninsula. Such introductions seem to mediate the naturalization of eucalypts, because once dense tree plantations are established, new seedlings can become ectomycorrhizal very rapidly through infection from the established fungal network. In many cases, the introduced eucalypts compete with native species, and the eradication of the eucalypts is difficult, especially in areas with a long history of large and extensive plantations. In Spain, the regional and national governments are promoting the eradication of eucalypts from natural and national parks.

An example is the National Park of Doñana. Doñana is well known for its variety of species of birds, either permanent residents, winter visitors from north and central Europe, or summer visitors from Africa, such as numerous types of geese and colourful colonies of flamingo. Doñana's configuration is the result of its past as the estuary of the Guadalquivir river, and mainly consists of beaches, coastal mobile dunes, marshes and lakes. Fauna in Doñana is rich and some in danger of extinction, such as the Iberian lynx, the Egyptian mongoose and the imperial eagle. The local wildlife of Doñana depends on the water level. Autumn rains brought life back to the marshes and filled the lagoons after the dry summer. Gradually, the water attains a uniform depth of 30–60 cm over vast areas, and the resulting marshes attract flocks of water birds of the most varied kind. Among other causes, the level of the freshwater of the marshes is in danger as a result of intensive plantations of eucalypts (Sacks et al. 1992). In the Doñana National Park (southern Spain), the eradication of eucalypts is necessary to reduce the water loss and the conservation of the local wildlife (e.g. Iberian lynx, *Felix pardina*) (Palomares et al. 1991).

The eucalypts introduced are competitive with native species, and their control and eradication are difficult. These eucalypts and the associated exotic ECM fungi can regenerate from root fragments. We know as well that ectomycorrhizal spores can remain dormant in soil for long periods, and might colonize eucalypt seedlings grown from remaining eucalypt seeds Moreover, ectomycorrhizal eucalypt seedlings often efficiently compete for soil nutrients with the planted young trees of native *Quercus* species (i.e. *Q. ilex* L., *Q. pyrenaica* Willd. and *Q. suber* L) (Muriel and Díez, unpublished). These reasons would account for the naturalization of eucalypts in Peninsular Spain (Vila et al. 2001) and in other regions of the Mediterranean Basin (Le Floc'h 1991).

In the checklist of alien species in Spain, Sanz-Elorza et al. (2001) classified the eucalypts as alien plants with a clear invasive behaviour, and as dangerous (causing ecological damage and alterations) for natural ecosystems. To date, the invasion behaviour is limited to areas with a long history of large and extensive plantations. In Spain, the red river gum often invades along the watercourse, which is its natural habitat in Australasia. In these areas, the plantations of eucalyptus have a great impact on lowering water tables, and have a devastating ecological impact, reducing soil quality and the habitat of local wildlife. The invasive tree species have a predictable set of life-history attributes, including low seed mass and short juvenile periods, and a short interval between seed crops (Rejmánek and Rhichardson 1996). Many eucalypts present many of these characteristics and produce large quantities of small seeds easily propagated by wind or/and animals (Pryor 1976). In addition, the pollination and seed dispersal do not limit the eucalypt invasiveness. Most eucalypts species are facultative outbreeders and are pollinated by a variety of generalist insects, in natural forest and exotic plantations (Pryor 1976). Therefore, the limited success of eucalypts as efficient invaders in the Iberian Peninsula can be puzzling. One can expect eucalypts to be successful as invaders, because these species are likely to be different in their resource utilization, easily escaping competition with native species (Richardson 1998).

To date, the difficulties of the eucalypts to find compatible fungal partners within the Iberian fungal biota seem to restrict their spread in the Iberian Peninsula to areas close to large plantations. The expansion of the invasion of the eucalypts from the plantation sites is likely to be hindered by the lack of compatible ectomycorrhizal fungi at potential seedling recruitment places. Lack of, or low extent of, colonization by compatible ectomycorrhizal fungi may be an important factor preventing or reducing seedling establishment of these alien trees. Eucalypt invasions would often result from the dispersal and propagules of Australian fungi. The importance of eucalypts as invaders correlates with the extent and the duration of planting, which relates to the dispersal rates of ECM propagules. The spread of propagules will facilitate a successful seedling establishment in new biotopes. As stressed above, the spread of ECM fungi in new biotopes is often by air spores, which is slow (Brundrett and Abbott 1995). This could explain why eucalypts might exist in plantations for many years, before they start to invade indigenous vegetation. A factor contributing to this lag is that compatible ectomycorrhizal propagules, in the form of spores (often air-born spores), needed time to accumulate in the soil (spore bank) before eucalypts can establish and proliferate. However, it is only a matter of time before the fungal spore bank reaches sufficiently high levels to allow ectomycorrhizal eucalypts to spread everywhere in compatible environments.

Invasiveness of Australasian ectomycorrhizal fungi

Awareness by politicians of the negative effects of exotic trees on natural biodiversity has led to an attempt to eradicate eucalypts in natural parks in Spain. But do Australasian ectomycorrhizal fungi persist after eucalypt eradication? After the removal of eucalypts, there will persist a high level of inoculums (spores, mycelium) of Australasian ectomycorrhizal fungi, which might colonize the roots of native trees that are planted on old eucalypt plantation sites. We have little information of the ability of the introduced Australian fungi to infect plants native to the Iberian Peninsula. It would be necessary to investigate whether these Australian fungi colonize roots of the Mediterranean trees planted on former eucalypt plantations. We do not know whether these exotic fungi spread beyond the plantations and colonize native ECM flora.

ECM fungi need to live in association with the tree roots, and their spread beyond the plantations will depend on (i) their compatibility with any native ectomycorrhizal plants, and (ii) their ability to exclude the indigenous ECM fungi. In our survey, we did not find Australian fungi in association with native ectomycorrhizal plants, except *Laccaria fraterna*, which naturally occurs in association with eucalypts in Australia and worldwide plantations. We found this Australian fungus in shrublands of the ectomycorrhizal shrub *Cistus ladanifer* L. near plantations of eucalypts. In two sites near the Monfragüe Natural Park, fruiting bodies of *L. fraterna* were found 500 and 700 m far from the nearest eucalypt tree, and no eucalypt seedlings were found near the fruiting bodies of *L. fraterna*. Molecular typing of the ectomycorrhizal root tips identified *L. fraterna* on roots of *Cistus ladanifer* (Muriel and Díez, unpublished). We do not know whether these native woody plants that are associated with exotic ectomycorrhizal fungi (i.e. *C. ladanifer*) have access to resources that these native plants normally cannot tap, which would modify the nutrient cycling and affect the ecosystem functioning.

In our study, we detected the Australian fungi *Urnula rhytidia* and *Discinella terrestis,* which are considered as saprophytic (or facultatively ectomycorrhizal). These fungi should have come with the eucalypts. *Urnula rhytidia* was also found on fallen leaves of *Quercus ilex* L., in oak woodlands near eucalypt plantations in Badajoz. These data suggest that many other saprophytic fungi could be introduced with the eucalypts and might spread to native ecosystems. Due to their saprophytic life style, these exotic fungi might persist for a long time after the eradication of the eucalypts, altering nutrient cycling in the soil.

Limiting further introduction of exotic EM

Australian forests have one of the most privative and rich mycobiota of the world (Castellano and Bougher 1994; Bougher and Syme 1998), but only a few Australian ECM fungi seem to have been introduced into the Iberian Peninsula. The limited functional and genetic diversity of introduced ECM species should be determining the environment range that eucalypts are able to invade. However, we continue to move soil and microbes around the world to establish new plantations, which favours the introduction of more and more Australian ECM fungi. Particular ECM fungal species might confer unique advantages for obtaining nutrients in potential habitats or to use particular nutrient resources (Buscot et al. 2000). An increased diversity of introduced Australian ECM fungi might increase the ability of the eucalypts to invade new habitats in the Iberian Peninsula.

Some authors propose to increase the diversity of the introduced ECM fungi to improve the productivity of exotic eucalypt plantations (Dell et al. 2002). This may involve selecting hardy

ECM strains tolerating a wide range of edaphic and environmental conditions and strains that elicit growth responses of eucalypt plantations (Neves-Machado 1995). Among these inoculums, there may be highly competitive strains, which could make the natural ecosystems more vulnerable to invasion by eucalypts. This would include strains coping with extremely harsh or toxic soil conditions. Rare and ecologically sensitive ecosystems, such as serpentine communities, may be particularly vulnerable to the introduction of Australasian ECM fungi with broad environmental tolerances.

Elevated levels of atmospheric carbon dioxide (global climate change) create an increasing concern. Exotic plantations are now promoted for their presumed capacity to provide a net sink of atmospheric carbon, and mycorrhizal fungi may play a critical role in terrestrial carbon exchange processes (Staddon et al. 2002). Chapela et al. (2001) described how exotic ectomycorrhizal fungi induce soil carbon depletion in pine plantations in Central America. We have no information on the impact the Australasian fungi could have on the cycles of nutrients in the Iberian soils.

Several methods for the transformation of ectomycorrhizal fungi are already available (Hanif et al. 2002; Pardo et al. 2002). Some scientists propose to use strains genetically manipulated to form better symbiotic systems, including ECM fungi which are more efficient in mobilizing nutrients from soils. This might result in eucalypt species that do not invade at present but become invasive if associated with such selected (or genetically modified) fungal strains.

Conclusions

Several conclusions can be drawn from our studies on the ectomycorrhizas of eucalypt plantations in the Iberian Peninsula:

(1) In the Iberian Peninsula, a number of Australasian ECM fungi were introduced together with the eucalypts. The reduced number of Australian ECM fungal species, together with the low ability of Iberian fungi to colonize eucalypt roots, would explain the low diversity of ECM fungal communities in exotic stands of eucalypts.

(2) The introduction of these Australasian ECM fungi appears to be one of the main factors accounting for the successful establishment of eucalypt plantations in the Iberian Peninsula, their naturalization, and invasive behaviour. Consequently, a deeper knowledge on the ectomycorrhizas of eucalypt stands in the Iberian plantations will help to refine our ability to predict the invasiveness of the eucalypts introduced. The knowledge generated may be crucial for determining potential endangerment and to suggest strategies for protecting the diversity of the Mediterranean ecosystems.

(3) It will be necessary to investigate potential host shifts of Australian fungi to native hosts, and their effects on the native ECM fungal communities and on the functioning of Mediterranean ecosystems. For these reasons, it is urgent to investigate the ectomycorrhizas of native trees planted in former plantation sites, and roots of indigenous ectomycorrhizal plants growing near eucalypt plantations.

(4) The present investigation highlights the need to regulate the translocation of ectomycorrhizal fungi for forest inoculations. Quarantine measures would be necessary to control any future introduction of ECM fungi of Australian origin.

(5) Before introducing beneficial Australasian strains of ECM fungi, we would recommend screening their ability to improve eucalypt invasiveness and to infect roots of native ECM plants. Screening the invasiveness of introduced strains will help to prevent negative effects on Iberian natural ecosystems.

Acknowledgements

The author is indebted to A. Muriel for her help during the fruiting body surveys, ectomycorrhiza sampling, and for her stimulating discussions. Special thanks are due to Dr J. Garbaye (INRA-Nancy, France) for critical comments on an early version of this manuscript. I sincerely thank Dr G. Moreno for his help with the identification of fruiting bodies of fungi. The assistance of Dr P. Rubio (Unit of Molecular Biology, University of Alcalá) with the DNA sequencing during the molecular identification of ectomycorrhizas is gratefully acknowledged. This research is supported by a 'Ramón y Cajal' researcher contract from the MCyT (Spain) (RC2002/2091), and a research grant funded by the University of Alcalá; 'Ectomycorrhizal fungi of eucalypt plantations in the Mediterranean – native or exotic fungi?' (Contract UAH2002/049).

References

Agerer R (1997) Descriptions of Ectomycorrhizae. Einhorn-Verlag Eduard Dietenberger GmbH Schwabisch Gmund, Munchen, Germany, 89 pp

Allen MF (1991) The Ecology of Mycorrhizae. Cambridge University Press, New York, 184 pp

Altschul SF, Madden TL, Schaffer AA, Zhang J, Zhang Z, Miller W and Lipman DJ (1997) Gapped BLAST and PSI-BLAST: a new generation of protein database search programs. Nucleic Acids Research 25: 3389–3402

Armstrong AJ and van Hensbergen HJ (1996) Impacts of afforestation with pines on assemblies of native biota in South Africa. South African Forestry Journal 175: 35–42

Binggeli P (1996) A taxonomic, biogeographical and ecological overview of invasive woody plants. Journal of Vegetation Science 7: 121–124

Bougher NL and Syme K (1998) Fungi of Southeastern Australia. University of Western Australia Press, Perth, 404 pp

Brundrett MC (1991) Mycorrhizas in natural ecosystems. In: Begon M, Fitter AH and Macfadyen A (eds) Advances in Ecological Research, Vol 21, pp 171–313. Academic Press, London

Brundrett MC and Abbott LK (1995) Mycorrhizal fungus propagules in the jarrah forest. II. Spatial variability in inoculum levels. New Phytologist 131: 461–469

Buscot F, Munch JC, Charcosset JY, Gardes M, Nehls U and Hampp R (2000) Recent advances in exploring physiology and biodiversity of ectomycorrhizas highlight the functioning of these symbioses in ecosystems. Fems Microbiology Reviews 24(5): 601–614

Cairney JWC (2002) *Pisolithus* death of the pan-global super fungus. New Phytologist 153: 199–201

Casadevall A and Perfect JR (1998) *Cryptococcus neoformans*. American Society for Microbiology Press, Washington, 549 pp

Castellano MA and Bougher NL (1994) Consideration of the taxonomy and biodiversity of Australian ectomycorrhizal fungi. Plant Soil 159: 37–46

Chakrabarti A, Jatana M, Kumar P, Chatha L, Kaushal A and Padhye AA (1997) Isolation of *Cryptococcus neoformans* var. *gattii* from *Eucalyptus camaldulensis* in India. Journal of Clinical Microbiology 35: 3340–3342

Chambers SM and Cairney JWG (1999) *Pisolithus*. In: Cairney JWG and Chambers SM (eds) Ectomycorrhizal Fungi. Key Genera in Profile, pp 1–31. Springer-Verlag, Berlin, Germany

Chapela IH, Osher LJ, Horton TR and Henn MR (2001) Ectomycorrhizal fungi introduced with exotic pine plantations induce soil carbon depletion. Soil Biology and Biochemistry 33: 1733–1740

Davis MR, Grace LJ and Horrell RF (1996) Conifer establishment in South Island High Country: influence of mycorrhizal inoculation, competition removal, fertilizer application, and animal exclusion during seedling establishment. New Zealand Journal of Forestry Science 26: 380–394

Dell B, Malajczuk N and Dunstan WA (2002) Persistence of some Australian *Pisolithus* species introduced into eucalypt plantations in China. Forest Ecology and Management 169: 271–281

Díez J (1998) Micorrizas del bosque mediterráneo. Reforestación, biotecnología forestal (micropropagación, micorrización 'in vitro') y ecología molecular. PhD Dissertation, Universidad de Alcalá, Alcalá de Henares, Spain, 242 pp [in Spanish]

Díez J, Anta B, Manjón JL and Honrubia M (2001) Genetic variability of *Pisolithus* isolates associated with native hosts and exotic eucalyptus in the western Mediterranean region. New Phytologist 149: 577–587

Dunstan WA, Dell B and Majajczuk N (1998) The diversity of ectomycorrhizal fungi associated with introduced *Pinus* spp. in the Southern Hemisphere, with particular reference to Western Australia. Mycorrhiza 8: 71–79

Egger KN (1995) Molecular analysis of ectomycorrhizal fungal communities. Canadian Journal of Botany 73 (Suppl 1): S1415–S1422

Eldridge K, Davidson J, Harwood C and van Wyk G (1994) Eucalypt Domestication and Breeding. Oxford University Press, New York, 320 pp

Fahey TJ (1992) Mycorrhizae and forest ecosystems. Mycorrhiza 1: 83–89

Gardes M and Bruns TD (1993) ITS primers with enhanced specificity for basidiomycetes: application to the identification of mycorrhizae and rusts. Molecular Ecology 2: 113–118

Gardes M and Bruns TD (1996) Community structure of ectomycorrhizal fungi in a *Pinus muricata* forest: above- and below-ground views. Canadian Journal of Botany 74: 1572–1583

Gardes M, White TJ, Fortin J, Bruns TD and Taylor JW (1991) Identification of indigenous and introduced symbiotic fungi in ectomycorrhizae by amplification of nuclear and mitochondrial ribosomal DNA. Canadian Journal of Botany 69: 180–190

Giachini AJ, Oliveira VL, Castellano MA and Trappe JM (2000) Ectomycorrhizal fungi in *Eucalyptus* and *Pinus* plantations in southern Brazil. Mycologia 92: 1166–1177

Glen M, Tommerup IC, Bougher NL and O'Brien PA (2001a) Specificity, sensibility and discrimination of primers for PCR-RFLP of larger basidiomycetes and their applicability to identification of ectomycorrhizal fungi in *Eucalyptus* forests and plantations. Mycological Research 105: 138–149

Glen M, Tommerup IC, Bougher NL and O'Brien PA (2001b) Interspecific and intraspecific variation of ectomycorrhizal fungi associated with *Eucalyptus* ecosystems as revealed by ribosomal DNA PCR-RFLP. Mycological Research 105: 843–858

Grove TS and Le Tacon F (1993) Mycorrhiza in plantation forestry. Advances in Plant Pathology 23: 191–227

Halling RE (2001) Ectomycorrhizal co-evolution, significance, and biogeography. Annals of the Missouri Botanical Garden 88: 5–13

Hanif M, Pardo AG, Gorfer M and Raudaskoski M (2002) T-DNA transformation in the ectomycorrhizal fungus *Suillus bovinus* using hybromicin B as a selectable marker. Current Genetics 41: 183–188

Hibbett DS, Gilbert LB and Donoghue MJ (2000) Evolutionary instability of ectomycorrhizal symbioses in basidiomycetes. Nature 407: 506–508

Higgins SI and Richardson DM (1998) Pine invasions in the southern hemisphere: modelling interactions between organism, environment and disturbance. Plant Ecology 135: 79–93

Horton TR and Bruns TD (2001) The molecular revolution in ectomycorrhizal ecology: peeking into the black-box. Molecular Ecology 10: 1855–1871

Ingleby K, Mason PA, Last FT and Fleming LV (1990) Identification of Ectomycorrhizae. HMSO, London, 112 pp

Le Floc'h E (1991) Invasive plants of the Mediterranean Basin. In: Groves RH and Di Castri F (eds) Biogeography of Mediterranean Invasions, pp 67–80. Cambridge University Press, Cambridge

Lu XH, Malajczuk N, Brundrett M and Dell B (1999) Fruiting of putative ectomycorrhizal fungi under blue gum (*Eucalyptus globulus*) plantations of different ages in Western Australia. Mycorrhiza 8: 255–261

Martin F, Díez J, Dell B, and Delaruelle C (2002) Phylogeography of the ectomycorrhizal *Pisolithus* species as inferred from the nuclear ribosomal DNA ITS sequences. New Phytologist 153: 345–358

Martín MP, Díez J and Manjón JL (2000) Methods used for studies in molecular ecology of ectomycorrhizae (Chapter 2.3). In: Martín MP (ed) Methods in Root–Soil Interactions Research – Protocols, pp 25–28. Slovenian Forestry Institute Press, Ljubljana

Molina RH, Massicotte H and Trappe J (1992) Specificity phenomena in mycorrhizal symbiosis: community-ecological consequences and practical implications. In: Allen MF (ed) Mycorrhizal Functioning, pp 357–423. Chapman & Hall, London

Neves-Machado MH (1995) La mycorhization contrôlée d'*Eucalyptus globulus* au Portugal et l'effet de la sécheresse sur la symbiose ectomycorhizienne chez cette essence. PhD Dissertation. Université Henri Poincaré, Nancy I, Nancy, France 149 pp [in French]

Newman EI and Reddell P (1987) The distribution of mycorrhizas among families of vascular plants. New Phytologist 106: 745–751

Onguene NA and Kuyper TW (2002) Importance of the ectomycorrhizal network for seedling survival and ectomycorrhiza formation in rain forests of south Cameroon. Mycorrhiza 12: 13–17

Palomares F, Rodríguez A, Laffitte R and Delibes M (1991) The status and distribution of the Iberian lynx, *Felis pardina* (Temminck), in the Coto Doñana area, SW Spain. Biological Conservation 57: 159–169

Pardo AG, Hanif M, Raudaskoski M and Gorfer M (2002) Genetic transformation of ectomycorrhizal fungi mediated by *Agrobacterium tumefaciens*. Mycological Research 106: 132–137

Parladé J, Álvarez IF and Pera J (1996) Ability of native ectomycorrhizal fungi from northern Spain to colonize Douglas-fir and other introduced conifers. Mycorrhiza 6: 51–55

Pera J, Álvarez IF, Rincón A and Parladé J (1999) Field performance in northern Spain of Douglas-fir seedlings inoculated with ectomycorrhizal fungi. Mycorrhiza 9: 77–84

Perry DA, Molina R and Amaranthus MP (1987) Mycorrhizae, mycorrhizospheres, and reforestation: current knowledge and research need. Canadian Journal of Forest Research 17: 929–940

Peterson RL and Bonfante P (1994) Comparative structure of vesicular–arbuscular mycorrhizas and ectomycorrhizas. Plant and Soil 159: 79–88

Pryor LD (1976) Biology of Eucalypts. Edward Arnold (Publishers) Ltd, London, 82 pp

Pryor LD (1991) Forest plantations and invasions in the mediterranean zones of Australia and South Africa. In: Groves RH and Di Castri F (eds) Biogeography of Mediterranean Invasions, pp 405–412. Cambridge University Press, Cambridge

Read DJ and Pérez-Moreno J (2003) Mycorrhizas and nutrient cycling in ecosystems – a journey towards relevance? New Phytologist 157: 475–492

Rejmánek M and Richardson DM (1996) What attributes make some plant species more invasive? Ecology 77: 1655–1661

Richardson DM (1998) Forestry trees as invasive aliens. Conservation Biology 12: 18–26

Richardson DM, Cowling RM and Le Maitre DC (1990) Assessing the risk of invasive success in *Pinus* and *Banksia* in Southern African mountain fynbos. Journal of Vegetation Science 1: 629–642

Richardson DM, Williams PA and Hobbs RJ (1994) Pine invasions in the Southern Hemisphere: determinants of spread and invadability. Journal of Biogeography 21: 511–527

Richardson DM, Allsopp N, ĎAntonio CM, Milton SJ and Rejmánek M (2000a) Plant invasions – the role of mutualisms. Biological Reviews of the Cambridge Philosophical Society 75: 65–93

Richardson DM, Pysek P, Rejmánek M, Barbour MG, Panetta FD and West CJ (2000b) Naturalization and invasion of alien plants: concepts and definitions. Diversity and Distribution 6: 93–107

Sacks LA, Herman JS, Konikow LF and Vela AL (1992) Seasonal dynamics of groundwater – lake interactions and Doñana National Park, Spain. Journal of Hydrology 136: 123–154

Sanz-Elorza M, Dana E and Sobrino E (2001) Checklist of invasive alien plants in Spain (Iberian Peninsula and Balearic Islands). Lazaroa 22: 121–131

Smith SE and Read DJ (1997) Mycorrhizal Symbiosis, 2nd ed. Academic Press, New York, 605 pp

Staddon PL, Heinemeyer A and Fitter AH (2002) Mycorrhizal and global environmental change: research at different scales. Plant and Soil 244: 253–261

Vila M, García-Berthou E, Sol D and Pino J (2001) Survey of the naturalized plants and vertebrates in peninsular Spain. Ecologia Mediterranea 27: 55–67

Warner P (1999) The CalEPPC list of exotic pest plants of greatest ecological concern in California (California Exotic Pest Plant Council). Retrieved from http://www.caleppc.org.org/documents/newsletter593.htm on 15 March 2004

White TJ, Bruns T, Lee SS and Taylor J (1990) Amplification and direct sequencing of fungal ribosomal RNA genes for phylogenetics. In: Innis MA, Gelfand DH, Sninsky JJ and White TJ (eds) PCR Protocols: a Guide to Methods and Applications, pp 315–322. Academic Press, New York

Zobel BJ, van Wyk G and Stahl P (1987) Growing Exotic Forests. John Wiley and Sons, New York, 508 pp

Biological Invasions (2005) 7: 17–27

Phenological and demographic behaviour of an exotic invasive weed in agroecosystems

Jordi Recasens*, Víctor Calvet, Alicia Cirujeda & Josep Antoni Conesa
*Departament Hortofructicultura, Botànica i Jardineria. ETSEA. Universitat Lleida Rovira Roure 191, 25198 Lleida, Spain; *Author for correspondence (e-mail: jrecasens@hbj.udl.es; fax: +34-973-238264)*

Received 4 June 2003; accepted in revised form 30 March 2004

Key words: *Abutilon theophrasti*, biology, invasiveness, maize, weed

Abstract

An experimental work was conducted in Lleida (Spain) aiming to characterise the phenology and to quantify the demographic processes regulating the populations of *Abutilon theophrasti* Medicus in maize fields. Seedling emergence started a few days after crop sowing in early May and continued during two more months. The vegetative phase was very long due to the late seeding emergence; these later emerged plants showed a slower development, and many of them did not reach the fertility stage. A flowering peak was observed 12 weeks after emergence in late July, and fruit dehiscence and seed setting started in mid August, several weeks before crop harvest. Four different cohorts were identified, and two main peaks of emergence were determined 21 and 49 days after crop sowing nearest related with field irrigation. A functional logarithmic relationship between cumulative growing degree-days (GDD) and cumulative emergence was also described. The resulting demographic diagram reflects greater values relating to seedling survival for May cohorts (90.2 vs 7.9%), to fertility (100 vs 75%) and to fecundity (3774 vs 92 seeds pl^{-1}) than those determined for the June cohorts. The late emerged plants are subjected to a high density and are strongly affected by light competition, and their reproductive phase initiation delay is of about 10–20 days. In an assay conducted in Petri dishes, the seeds provided from plants emerged earlier were found more vigorous and germinated more than those from late emerged plants, which seem to be affected by incomplete fruit and seed ripening. Following the crop cycle without any weed control, the population rate increase was about 21.2. These values explain the high invasion capacity of this weed in the local summer irrigated fields, which consists in assuring their presence through a persistent soil seed bank and increasing the probability to spread to other fields.

Introduction

Most of the weeds in arable crops show a clear adaptation to the cultural techniques. This enables them to tolerate the proposed control strategies such as tillage, mowing, crop rotations, fallow, herbicides. The efficacy of any of these techniques therefore requires a detailed knowledge of the weed species' biology, especially referring to the phenological and demographic behaviour. The detection of the critical cycle phase, during which the main demographic regulation takes place, allows the identification of the most appropriate moment for the control method. This also allows the estimation of the expected economic benefit of weed control at short term taking into account the costs of these techniques and the infestation degree of the weed species. Each infestation situation together with the expected objectives define the correct control strategy: prevention, contention, reduction or eradication (Fernández Quintanilla 1992). The main aspects affecting this decision are on the one hand the economic value of the crop and on the other hand the biological behaviour of the weed. The assessment of this latter aspect is generally complex as besides the degree of harm caused by the weed species other phenologial and demographic parameters are considered. The definition of a control strategy of an exotic weed is urgent and necessary but presents an additional difficulty due to the lack of information of the biological behaviour in the new geographical area and agroecosystem. A clear example of this situation occurred with the presence and expansion of *Abutilon theophrasti* Medicus (velvetleaf) during the 1980s in the irrigated area of Aragon and Catalonia (northeastern Spain) (Zaragoza 1982; Izquierdo 1986). Even if its presence in Spain, more concrete, in Catalonia, had been quoted several times since the XIX century as a transitory weed (Casasayas and Masalles 1981; Casasayas 1989), its invasive behaviour was not detected until the 1980s when the weed was introduced infesting maize, sorghum and soyabean imported from the USA.

This invasive process was also observed more or less at the same time in Holland, Italy and France (Häfliger 1979; Cantele et al. 1987). In Spain, the presence of *A. theophrasti* as an infesting weed started in the irrigated area of the Urgell (Lleida) where it established itself with success in the maize fields showing a great capacity of adaptation and competition. The economic losses in these first years were especially important not only due to the damage caused by the weed but also due to the lack of selective herbicides to control it (Izquierdo 1986). In 1991, around 10,000 ha of maize were affected corresponding to 44% of the cultivated maize area. This shows that the expansion is up to sixfold higher than the one observed seven years before (Calvet and Recasens 1993).

Spreading of *A. theophrasti* in crop fields in Catalonia presumably occurs through the joint harvest of weed seeds with the corn, its posterior incorporation in clamp, which is then spread as a caw manure on other fields (Izquierdo 1986). The survival capacity of these seeds in maize silage and in slurry is well known, and a similar weed spread mechanism was observed in Holland (Elema et al. 1990; Bloemhard et al. 1992).

A. theophrasti is a summer annual C_3 weed with rapid growth and high photosynthetic rates (Regnier et al. 1988). Control is difficult due to the extremely long-living seeds (Egley and Chander 1983; Lueschen et al. 1993), which exhibit hardseededness (Horowitz and Taylorson 1984). Velvetleaf is an excellent competitor due to its rapid root growth compared to other weed species and due to its allelopathic effects on crop plants (Warwick and Black 1986). The species exhibits characteristic genetic traits of colonizers, including polyploidy, self-fertilization, and high levels of population differentiation (Warwick 1990; Warwick and Black 1986).

A better understanding of the phenological and demographic behaviour of *A. theophrasti* weed populations in Mediterranean crop fields is needed to optimise the efficacy of a control program. The objectives of this research were to characterise the phenology and demography of an *A. theophrasti* population, including the initial time of emergence, to describe the demography of different cohorts, to find the percentage of germinability of their seeds after production and to calculate the rate of increase of the weed population during a cropping season.

Materials and methods

Site description of the experimental field and general procedures

Field experiments were conducted during 1998, in a 2 ha commercial maize monocrop at El Poal (Lleida, Spain) where a natural infestation of velvetleaf was detected during the last few years. The soil was described as a Fluventic Xerochrepts with 2.9% organic matter and pH 8. The experimental field was chisel plowed in early April prior to the maize sowing, which took place on 26 April. No herbicides were applied in the experimental field plot during the trial. The experimental design consisted of a 10 × 15 m grid, within 15 permanent 1.5 × 1 m squares were randomly selected for the phenological and demographic estimations.

Daily rainfall and maximum and minimum air temperatures were obtained from a meteorology station less than 1 km away from the experimental field. Growing degree days (GDD) were calculated as the mean of the daily minimum and maximum air temperatures minus the 10 °C base temperature (Gupta 1985; Benvenuti and Macchia 1993).

Phenology

During the crop cycle, the different phenological stages of all the *A. theophrasti* plants found in the different selected squares, were estimated following the BCCH scale (Hess et al. 1997).

Seed bank estimates

Two soil sampling dates were considered for the seed bank study. One was sampled shortly after the crop sowing and the other after the crop harvest. In each case, five soil cores measuring 5 cm in diameter and 20 cm in depth, were collected in a W-shaped pattern from each permanent square. The total soil surface area sampled was 98.72 cm^2 per quadrat. Following Dessaint et al. (1990) and Mulugeta and Stoltenberg (1997), soil core samples were stored for 2 weeks in the PVC cores at 4 °C with the aim to prevent seed germination. Later, the samples were water-cleaned and the organic fraction, containing velvetleaf seeds, placed in aluminium trays in the greenhouse. The samples were surface watered whenever necessary. Weed seedling emergence in the greenhouse was quantified periodically. Three months later, the dormant seeds remaining in the soil samples were identified and separated using a pair of binoculars. During the identification, seeds, which remained firm when pressed by fine tipped nippers, were considered viable. The amount of dormant seeds and greenhouse seedlings were summed up to obtain the viable seed count. The seed number was expressed on the basis of soil surface.

Over-ground demography

Velvetleaf seedling emergence in each square was determined by regular assessments and marking newly emerged seedlings with a coloured wire. The mean time of emergence (MTE) was calculated following the methods of Mulugeta and Stontelberg (1998), where n_i was the number of seedlings at time i and d_i was the number of days from day 0 of the experiment (the crop sowing) to time i.

$$\mathrm{MTE} = \sum n_i d_i \Big/ \sum n_i$$

A differently coloured wire was used on each approximately weekly sampling date. Velvetleaf plants marked with the same colour belonged to the same cohort. The different initial times of seedling emergence (16 and 26 May, 13 and 20 June) were referred to as cohorts (A1, A2, A3, A4), respectively. The time of emergence was established from the soil disturbance, which happened at the crop sowing. Weed seedlings that had emerged before the crop sowing were removed by hand. Plant density counts were followed until crop harvest, and velvetleaf seed production was measured by hand-harvesting seed capsules from all plants of each quadrat as they matured.

Seed germination in Petri dishes

To estimate the seed germinability, seeds of velvetleaf were collected from the A1 and A2 cohorts, only, as insufficient seeds were obtained from the other two cohorts. Twenty seeds of each cohort were placed on filter paper in a 9 cm diameter plastic Petri dish. The seeds were watered daily with 4 ml of distilled water and maintained in a growth chamber at 12–24 °C with a 12 h photoperiod. Germination of seeds with a protruded radicle of seed length or greater were counted under a fluorescent lamp. Seed germination was determined daily for two weeks after exposure to water. Germinated seeds were removed from the Petri dishes with nippers and then counted. The experiment was conducted in a completely randomised design with 20 replications.

Statistical analysis

Cumulative emergence data of velvetleaf were analysed by least-squares regression. The functional relationship between cumulative emergence and cumulative GDD was determined. Data of soil seed bank, seedling emergence on quadrats and seed germination in Petri dishes were analysed using the ANOVA procedure, and the differences between soil sampling time or cohorts were determined using the Fisher Protected LSD test at a 5% level of significance. For the statistical analysis, the SAS program was used (SAS 1999).

Results and discussion

Phenology

A. theophrasti seedling emergence started in less than two weeks after crop sowing (early May) and continued for two months until mid June (Table 1). The vegetative development phase characterized by the appearance of leaves and buds, stem elongation and of the lateral branches is normally very long so that a low percentage of plants were still at this stage even in mid September. This prolongation of the vegetative phase is due to the presence of plants, which survive until the end of summer without reaching the fertility stage. The vegetative phase lasted a little longer than two months for the first germinating plants in May even if it was not rare to observe still a low percentage of vegetative plants in August. Flowering started at the beginning of June and continued until mid-September. A flowering peak occurred 12 weeks after emergence corresponding to late July even if this maximum was not very evident due to the long lasting flowering phase. However, maximum flowering percentages were observed between mid July and mid August. Fruit dehiscence and seed setting started in mid August and continued during autumn until maize harvest. At this moment, a bit more than half of the population was already dying off. This phenological gradient means that during the summer, plants at most of the development stages can be found within the same population.

The phenological behaviour of this weed species in the present trial has been found to be similar to the one observed in the fields of southern and central USA (Hartzler et al. 1999) but a bit earlier than the one observed in eastern Canada and northern USA (Warwick and Black 1988) where the flowering starts from late August to September and seed setting occurs from September to October. On the other hand, compared with the observation made in Italy (Cantele et al. 1987; Viggiani 1995), the present population shows a slight phenological delay, especially referring to the beginning of emergence. These

Table 1. Percentage of plants of *Abutilon theophrasti* at each phenological stage during the crop season.

Date	d.a.s[b]	Phenological stages[a]							
		10	20	30	40	50–60	70	80	90
26 April	0								
9 May	13	100							
16 May	20	98	2						
23 May	27	81	19						
30 May	34	3	96	1					
6 June	41	1	52	47					
13 June	48	41	1	58					
20 June	55	20	26	54					
1 July	66		40	49	11				
5 July	70		37	50	9	3	1		
13 July	78		33	34	19	9	4		
20 July	85		28	27	2	15	8		
27 July	92		15	28	23	18	16		
3 August	99		5	24	26	13	32		
7 August	103		3	24	17	14	42		
14 August	110		1	25	11	19	44		
21 August	117		1	10	8	15	50	16	
31 August	127			5	4	6	35	49	
7 September	134			3	2	3	27	65	
14 September	141			1	1	3	17	78	
21 September	148					3	7	86	4
30 September	157						6	38	56

[a] Phenological stages: 0, germination; 10, emergence; 20, leaf and shoot development; 30, stem elongation; 40, side shoot development; 50, flowerbud development; 60, flowering; 70, fruit development; 80, seed maturity; 90, dying off.
[b] Days after crop sowing.

geographical variations in the phenological behaviour of *A. theophrasti* can be considered as normal due to the climatic variability of the compared environments on the one hand, and on the other hand, the type of crop and variety can be sown at different moments, which can oscillate within more than 4 weeks inside the same region. Besides the presence of this weed species in other Spanish areas, especially in maize and cotton fields of southern Spain (Andalucía) (Cortés et al. 1998), no phenological data are available from these populations, which would allow the analysis of possible differences between them.

In this species, flowering and seed production showed a short-day photoperiodic response (Sato et al. 1994), and its competitive ability could be influenced by the daylength. However, Steinmaus and Norris (2002) indicated that velvetleaf had a range of growth responses to a variety of light availabilities and that it should have little difficulties in becoming fully established in the irrigated agroecosystems of Mediterranean-type regions. From a climate-matching approach, it appears that the Mediterranean climate is a deterrent to the integration of velvetleaf, and that its persistence in the same Mediterranean regions is more closely linked to the use of irrigation in agriculture than to climatic factors (Holt and Boose 2000).

Seedling emergence

Initial *A. theophrasti* emergence occurred at 18 DAS (days after sowing) at the same time as the maize emergence, and reached a first peak at 21 DAS with 53% of the total emergence (Figure 1). Following 3 weeks without emergence, a second emergence event corresponding to approximately 38% of the cumulative emergence occurred 49 DAS, a few days after irrigation. A maximum seedling density of 14 plants m^{-2} was observed, 56 days after crop sowing. The emergence is well identified by four different cohorts (Table 2). Plant densities of A1 and A3 cohorts were, respectively, 6.6 and 5.8 seedlings m^{-2} and were greater than that of the other cohorts (A2 and A4) which were only 1.2 and 1.0 seedlings m^{-2} respectively as estimated.

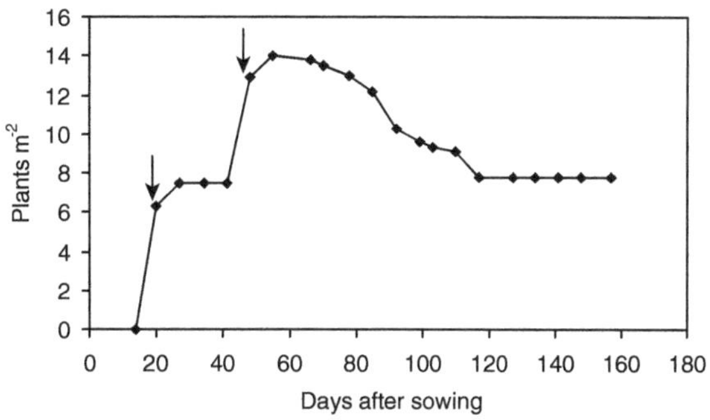

Figure 1. Changes in the total *Abutilon theophrasti* density during the crop cycle. The arrows indicate the field irrigation dates.

Similar dates were observed for populations growing in the USA (Cardina and Norquay 1997; Harztler et al. 1999) and Italy (Cantele et al. 1987), with two emergence peaks nearest related with soil humidity caused by rain or field irrigation. Other secondary factors, such as the soil ploughing similar to the one conducted previously in our field, specifically influence the earlier velvetleaf emergence (Mohler and Galdorf 1997). Exposure of seeds to light, improved soil aeration, increased loss of volatile inhibitors from soil and movement of seeds to more favourable germination sites have been suggested as possible causes of increased seed germination and emergence of seedlings (Egley 1986).

A mean time of emergence (MTE) value of 41.4 was estimated, which fits into the range

Table 2. Plant density and cumulative growing degree days (GDD) of four *Abutilon theophrasti* cohorts.

Cohort	Initial time of emergence	Cumulative GDD from crop sowing to weed emergence	Days from crop sowing to emergence time	Plant density at emergence time Pl m^{-2}	Significance[a]
A1	Mid-May (16/5)	120.9	21	6.6	a
A2	Late May (26/5)	204.0	31	1.2	b
A3	Mid-June (13/6)	321.7	49	5.7	a
A4	Late June (20/6)	378.8	56	1.0	b

[a] Values with different letters are different for $P \leq 0.05$.

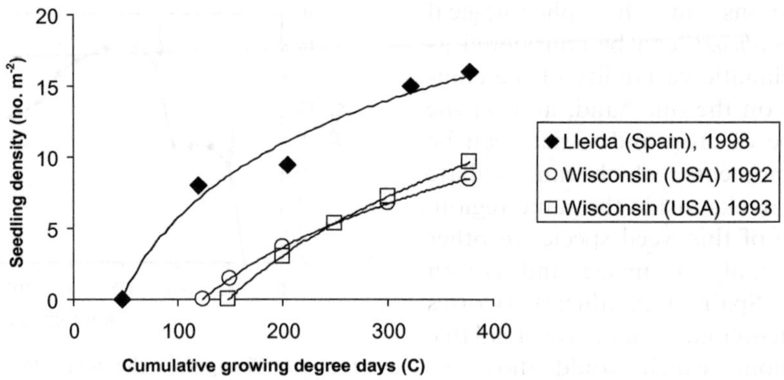

Figure 2. The influence of growing degree days (GDD, base temperature 10 °C) on cumulative seedling density of *Abutilon theophrasti* in Lleida (Spain) in comparison with those estimated in Wisconsin (USA) (Mulugeta and Stoltenberg 1997). Cumulative emergence was described by $y = 7.519 \ln x - 28.892$.

(from 30 to 42) observed by Hartzler et al. (1999) for this species during three different cropping seasons.

Pattern emergence

The functional relationship between cumulative GDD and cumulative emergence was described by a logarithmic model (Figure 2). A similar logarithmic function was established by Mulugeta and Stoltenberg (1997), but our data showed an earlier and greater emergence for the same cumulated GDD in comparison with those American populations. In all cases, emergence was nearly complete after 380 GDD.

In accordance with these authors, the emergence length of this species appeared to be affected by GDD accumulation and rainfall. In our region, rainfall is very low (<400 mmm year^{-1}), and the summer is the typically driest season of the Mediterranean climate. The higher pattern of emergence of *A. theophrasti* observed in our fields seems to be promoted by the periodic field irrigation which brings the humidity needed for seeds that were in the soil.

Seedling survival

Data on velvetleaf seedling cohort survival are shown in Table 3. In the first cohort (A1), which showed a greater seedling density, a 50% mortality was observed. This cohort showed a greater plant density at crop harvest (4 plants m^{-2}). The greater survival (74%) of cohort A2 does not correspond to a significant adult plant density because a small number of seedlings emerged. Only 10% of the plants from cohort A3 reached maturity, and there was no plant survival was detected from the A4 cohort. At harvest time, the final plant density was mainly represented by the first cohort A1 with a 71%. These data show that the survival of velvetleaf in a maize field depended on the seedling emergence moment. A first approach of these results indicated that this survival reduction may be due to interspecific and intraspecific competition. This effect is magnified in the June cohorts, which had more difficult conditions to survive compared to more developed plants. Lindquist (1995) suggested a competition for light so that the late velvetleaf seedlings would be more affected due to a great canopy closure than to the present plant density.

Table 3. Plant density and seed production of four *Abutilon theophrasti* cohorts before crop harvest time.

Cohort	Cumulative GDD between crop sowing to crop harvest	Seedling survival (%)	Plan density at harvest (Pl m^{-2})	Seed production (seeds plant^{-1})	Seed production (seeds m^{-2})
A1	1506.5	50	4.0	5402	513,190
A2	1400.8	74	1.1	2177	43,540
A3	1283.1	10	0.5	92	7452
A4	1225.8	0	0.0	0	0

Seed production

Velvetleaf seed production per plant differs between A1 (5402 seeds plant^{-1}) and A2 (2177 seeds plant^{-1}) and was greater than the production at A3 (92 seeds plant^{-1}) (Table 3). Seed production per unit area did no differ among A1 (513190 seeds m^{-2}) and A2 (43540 seeds m^{-2}). By comparison, Zanin and Sattin (1988) observed an average velvetleaf seed production of 3379 seeds plant^{-1} for Italian populations when grown together with maize. Warwick and Black (1988) indicated a variable fecundity of this species ranging from 700 to 17,000 seeds depending on their origin, field crop or plant size.

As was noted, for this species, by Lindquist et al. (1995), the differences between cohort fecundity would be due to the fact that the earlier emerging plants will have an inherent advantage over later emerging individuals and therefore will be responsible for producing a greater number of seeds. These late emerged plants in a high density of other plants seem to present delayed reproduction and show a lower allocation to reproduction due to competition for light (Mabry and Wayne 1997). Under shady conditions, the plants can show a suppressed growth and seed production (Bello et al. 1995) or a time delay of about 10–20 days for the initiation of reproductive allocations (Steinmaus and Norris 2002).

Seed bank population

There was a large difference between the number of velvetleaf seeds recovered in October (after crop harvest) and the number of seeds found in the seed bank immediately before crop sowing (Table 4). The seed bank at the beginning of the trial had 215 seed m^{-2}. We do not know the amount of surviving seeds during the cropping cycle, but this figure should not be more than 200 seeds m^{-2}, corresponding to the value of the initial seed bank minus total emerged seeds. However, emergence is only one source of the seed loss from the seed bank and predation, disease, fatal germination corresponding to germination without emergence and seed death should be added. After the seed rain and crop harvest, the seed bank was of 4567 seeds m^{-2}. This value corresponds to a 21-fold increase of the initial seed bank. Despite the fact that these data vary from the trial to the commercial field because no control was done in the experiment, they reflect a part of the species' cycle and specifically the population increase that can be expected. The low initial seed bank density suggests that the present field had been recently infested with *A. theophrasti*. The amount found is similar to that of the velvetleaf seed bank estimated by Cardina and Norquay (1997) in a maize field at a 20 cm depth where only a single year seed production was allowed.

Table 4. Soil seed density of *Abutilon theophrasti* before crop sowing and after crop harvest.

	Seeds m^{-2}	Significance[a]
Before crop sowing	215	a
After crop harvest	4567	b

[a] Values with different letters are different for $P \leq 0.05$.

The percentage of seeds in the seed bank that became seedlings, i.e. the emergence rate was 6.5% (data not shown). This value is very similar to the one reported by Lindquist et al. (1995) and 2% higher than those estimated by Pacala and Silander in soybean (1990). This similarity reflects that, in the same season, the emergence of a cohort from the seed bank is independent of the number of previous cohorts, which have previously germinated and seems to be mainly promoted by the field irrigation during the late spring season.

Velvetleaf emergence increases with shallow burial and decreases at greater burial depths (Cantele et al. 1987; Mohler and Galdorf 1997). Although temperature and soil moisture conditions are probably not limiting, germination is rare in seeds buried at more than 10 cm depth (Cardina and Sparrow 1997) so that then an important source of seeds remain dormant. The main fraction of the soil seed bank estimated in October is derived from the seed rain shed by mature plants. During the next season, depending on the soil tillage and on the depth in which the seeds will be placed, the survived seeds will be able to germinate or will continue being dormant. In all cases, a permanent seed bank has been formed, and the weed infestation is guaranteed for several years.

Demographic diagrammatic model

As an annual weed, five demographic processes regulate the population dynamics of *A. theophrasti*: seedling recruitment and survival, seed production, seed dispersal, and seed survival in the soil. A schematic diagram of the annual population dynamics is shown in Figure 3. The boxes represent state variables that can be measured in the field. The valve symbols represent the five demographic processes, each of which may be influenced by a number of factors including competition, predation and migration. In Figure 3, separated data are shown for the two main flux seedling emergence observed: May cohort corresponding to A1 + A2 and June cohort corresponding to A3 + A4. For both monthly cohorts, similar rates of emergence of 3.6% and 3.2%, respectively, were observed, but great differences in seedling survival were registered: 90.2% in the May cohort and 7.9% in the June cohort. These differences are also reflected in the different plant densities. The mortality causes of the youngest cohort have been commented above and are mainly promoted by an interspecific and intraspecific competition for light. In the June cohort, it was also observed that several plants did not reach the fertility stage. The seed production was also very different for the May and June cohorts: 3774 seeds plant^{-1} and only 92 seeds plant^{-1}, respectively. No data on seed exportation with the harvest machinery or about depredation by animals were obtained. The rate of population increase during the crop cycle (Δ) was calculated as the coefficient between both soil seed bank estimates. This value is very high (>21) and shows a vigorous capacity of this species to spread in a new habitat. The absence of herbicide application and the absence of any other weeds in this experience reflect the behaviour of this species in the earlier years of their introduction in Spanish maize fields being capable to compete with the crop and to replace other weeds.

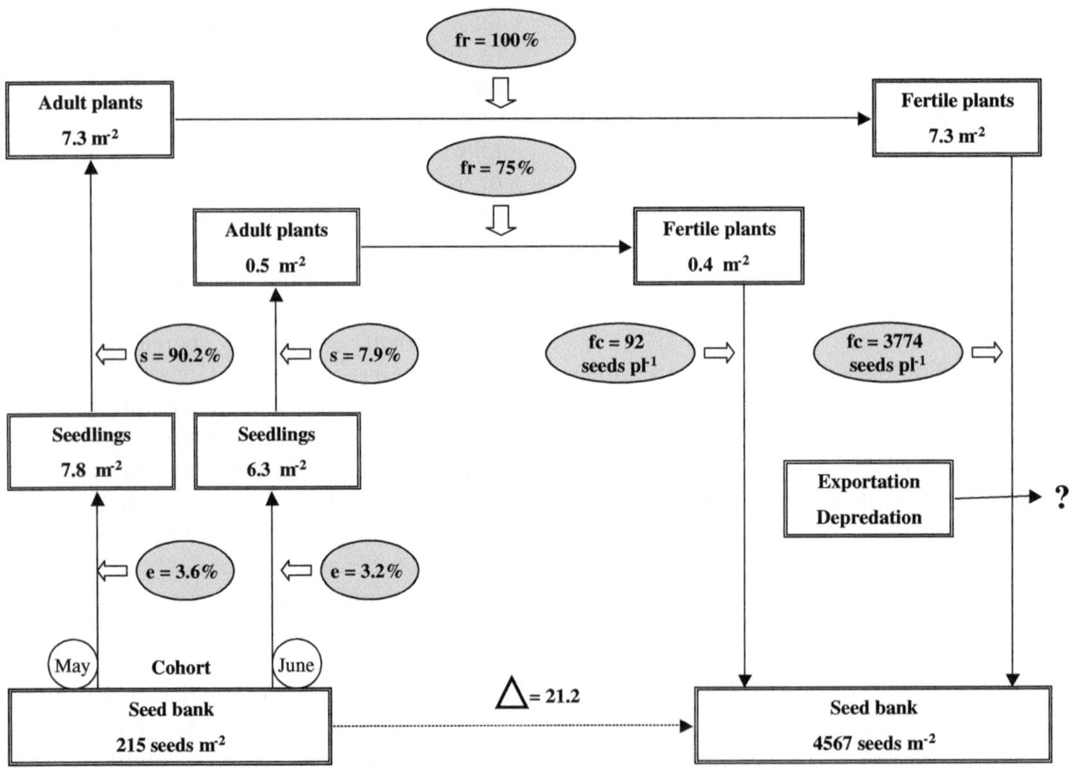

Figure 3. Demographic diagrammatic model for both monthly cohorts (May cohort and June cohort) of *Abutilon theophrasti* population present in a maize crop. e, emergence; s, seedling survival; fr, fertility; fc, fecundity; Δ, rate of population increase during the crop cycle (April–October).

Table 5. Germinability of harvested seeds of *Abutilon theophrasti* for the two earlier emerged cohorts.

Cohort	Emergence time	Germination (%)	Significance[a]
A1	Mid-May (16/5)	81.5	a
A2	Late May (26/5)	38.8	b

[a] Values with different letters are different for $P \leq 0.05$.

Seed germinability

A significant difference between the A1 and A2 cohorts on seed germinability in Petri dishes in the laboratory was found (Table 5). Big differences in the percentage of germinability of seeds from the A1 and A2 cohorts were found (81% and 37%, respectively). These results are in concordance with those of Sato et al. (1994), who noted that seeds of this species produced by earlier germinated plants were heavier and more vigorous than those which germinated later. Two causes can explain these differences: (1) a higher dormancy appears in the seeds produced by late germinated plants or (2) incomplete fruit and seed ripening in the plants occurs in these later cohorts. Velvetleaf shows a typically primary dormancy known as 'hardseededness', which is caused by the seed coat being impermeable to water (Warwick and Black 1986). This seed dormancy is shortened in shaded plants (Bello et al. 1995), but it is unclear whether the decrease in seed dormancy of shaded plants does result from the fact that few seeds have a hard seed coat or if some other dormancy mechanisms are responsible for it. If we understand that our A2 cohort is composed mainly of shaded plants, our results are not in concordance with the expected decrease in dormancy. A possible cause of these differences would be the delayed allocation observed in the late germinating plants and, as a consequence, an incomplete fruit and seed maturity. Steinmaus and Norris (2002) confirm, for this species, a time delay of about 10–20 days for the initiation of reproductive allocation in shaded plants in comparison with plants grown in full sunlight.

Implications for management

The phenological behaviour of *A. theophrasti* in our maize fields do not differ substantially from those registered in Italy or USA by other authors. The extended period of the different phenological stages, and specially the possibility to emerge over a long period of time and their ability to shed seeds several weeks before crop harvest, help to characterize their high phenotypic plasticity. The emergence time of *A. theophrasti* appeared to be affected by GDD accumulation and rainfall (Hartzler et al. 1999). In Mediterranean maize crops, the irrigation replaces the rainfall as a factor favouring the seedling emergence. Two main cohorts were well identified in concordance with time of two relevant agricultural practices: the crop sowing and the next irrigation applied to the field approximately 1 month later. Although other later emergences would be possible, only both these earlier cohorts are determinants in the persistence of the population in further years. The high fecundity levels that have been estimated in this species and the known high longevity of their seeds (Zanin and Sattin 1988) promote a 'persistent seed bank' as was defined by Thompson and Grime (1979). *A theophrasti* is in general more of a problem in monocultures, where one crop is grown year after year, than in rotational sequences, because it has a better chance of building up a seed bank in the soil.

Diverse practices are proposed to sustain effective control of *A. theophrasti* infestations, including crop rotation, multiple herbicide applications and soil tillage (Warwick and Black 1988), but their efficiency is always incomplete. Once the species becomes established in a field, it is very difficult to eradicate it, and the use of density-based weed threshold treatments may not be an appropriate or effective weed control strategy for long term management of the species in maize (Cardina and Norquay 1997). Preventive practices to reduce the chances of its introduction into an infested field should be the best strategy. This prevention will require thorough cleaning of farm equipment that has passed through infested fields. In addition, due to the spread of *A. theophrasti* in our summer irrigated crops with the caw manure, a high level of attention is required for the maize silage used as animal food. Farmers must refuse this forage if the presence of *A. theophrasti* seeds is confirmed.

Further research is needed on the control strategies of this weed. The cost and the difficulty to establish a threshold herbicide treatment, obligated to enhance the prevention practices if their introduction in the field would be possible, or apply an efficient eradication programme, hand weeding included, if this species is already present in the field.

References

Bello IA, Owen MDK and Hatterman-Valenti HM (1995) Effect of shade on velvetleaf (*Abutilon theophrasti*) growth, seed production, and dormancy. Weed Technology 9: 452–455

Benvenuti S and Macchia M (1993) Calculation of threshold temperature for the development of various weeds. Agricoltura Mediterranea 123(3): 252–256

Bloemhard CMJ, Arts MWMF, Scheepens PC and Elema AG (1992) Thermal inactivation of weed seeds and tubers during drying of pig manure. Netherlands Journal of Agricultural Science 40(1): 11–19

Calvet V and Recasens J (1993) Caracterización fenológica y demográfica de *Abutilon theophrasti* Medicus como mala hierba introducida en el cultivo del maíz en Lleida. Proceedings of the Congreso de la Sociedad Española de Malherbología, pp 93–97, Lugo

Cantele A, Zanin G and Zuin MC (1987) *Abutilon theophrasti* Medicus: II. Scalarità di emergenza dei semi nel terreno. Rivista di Agronomia 21(3): 221–228

Cardina J and Norquay HM (1997) Seed production and seed bank dynamics in subthreshold velvetleaf (*Abutilon theophrasti*) populations. Weed Science 45: 85–90

Cardina J and Sparrow DH (1997) Temporal changes in velvetleaf (*Abutilon theophrasti*) seed dormancy. Weed Science 45: 61–66

Casasayas T (1989) La flora al.lòctona de Catalunya. Catàleg raonat de les plantes vasculars exòtiques que creixen sense cultiu al NE de la Península Ibérica. PhD Thesis. Departament de Biologia Vegetal, Facultat de Biologia, Universitat de Barcelona: 429–431

Casasayas T and Masalles R (1981) Notes sobre la flora al.lòctona. Butlletí Institució Catalana d'Història Natural 46 (sec. Bot. 4): 111–115

Cortés JA, Castejón M and Mendiola MA (1998) Incidencias del *Abutilon* en el valle del Guadalquivir. Agricultura 796: 924–927

Dessaint F, Barralis G, Beuret E, Caixinhas ML, Post BJ and Zanin G (1990) Étude coopérative EWRS: la détermination du potentiel semencier I: recherche d'une relation entre la moyenne et la variance d'échantillonage. Weed Research 30: 421–429

Egley GH (1986) Stimulation of weed seed germination in soil. Review Weed Science 2: 67–89

Egley GH and Chandler JM (1983) Longevity of weed seeds after 5.5 years in the Stoneville 50-year buried-seed study. Weed Science 31: 264–270

Elema AG, Bloemhard CMJ and Scheepens PC (1990) Risk-analysis for the dissemination of weeds by liquid manure. Mededelingen van de Faculteit Landbouwwetenschappen 55(36): 1203–1208

Fernández Quintanilla C (1992) Bases para el control de las malas hierbas en sistemas de agricultura sostenible. Información Técnica Económica Agraria 88V(3): 143–152

Gupta SC (1985) Predicting corn planting dates for moldboard and no tillage systems in the corn belt. Agronomy Journal 77: 446–455

Häfliger E (1979) Grass weeds, a world wide problem in maize crops. Maize monograph Ciba Geigy 33–37

Hartzler RG, Buhler DD and Stoltenberg DE (1999) Emergence characteristics of four annual weed species. Weed Science 47: 578–584

Hess M, Barralis G, Bleiholder H, Buhr L, Eggers TH, Hack H and Stauss R (1997) Use the extended BBCH scale – general for the descriptions of the growth stages of mono- and dicotyledonous weed species. Weed Research 37: 433–441

Holt JS and Boose AB (2000) Potential for spread of *Abutilon theophrasti* in California. Weed Science 48: 43–52

Horowitz M and Taylorson RB (1984) Harseededness and germinability of velvetleaf (*Abutilon theophrasti*) as affected by temperature and moisture. Weed Science 32: 111–115

Izquierdo J (1986) Algunas características de *Abutilon theophrasti* Medicus, como mala hierba introducida en Lérida. Información Técnica Económica Agraria 65: 45–55

Lindquist JL, Maxwell BD, Buhler DD and Gunsolus JL (1995) Velvetleaf (*Abutilon theophrasti*) recruitment, survival, seed production, and interference in soybean (*Glycine max*). Weed Science 43: 226–232

Lueschen WE, Andersen RN, Hoverstad TR and Kanne BK (1993) Seventeen years of cropping systems and tillage affect velvetleaf (*Abutilon theophrasti*) seed longevity. Weed Science 41: 82–86

Mabry CM and Wayne PW (1997) Defoliation of the annual herb *Abutilon theophrasti*: mechanism underlaying reproductive compensation. Oecologia 111: 225–232

Mohler CL and Galford AE (1997) Weed seedling emergence and seed survival: separating the effects of seed position and soil modification by tillage. Weed Research 37: 147–155

Mulugeta D and Stoltenberg DE (1997) Seed bank characterization and emergence of a weed community in a molboard plow system. Weed Science 45: 54–60

Mulugeta D and Stoltenberg DE (1998) Influence of cohorts on *Chenopodium album* demography. Weed Science 46: 65–70

Pacala SW and Silander JA (1990) Field test of neighborhood population dynamics models of two annual weed species. Ecology Monograph 60: 113–134

Regnier EE, Salvuci ME and Stoller EW (1988) Photosynthesis and growth responses to irradiance in soyabean (*Glycine max*) and three broadleaf weeds. Weed Science 36: 487–496

Sato S, Tateno K and Kobayashi R (1994) Influence of seeding date on flowering and seed production of velvetleaf (*Abutilon theophrasti* Medicus). Weed Research Japan 39: 243–248

SAS (Statistical Analysis Systems) (1999) SAS/STAT User's Guide. Cary, North Carolina: Statistical Analysis Systems Institute, 1028 pp

Steinmaus SJ and Norris RF (2002) Growth analysis and canopy architecture of velvetleaf grown under light conditions representative of irrigated Mediterranean-type agroecosystems. Weed Science 50: 42–53

Thompson K and Grime JP (1979) Seasonal variation in the seed bank of herbaceous species in ten contrasting habitats. Journal of Ecology 67: 893–921

Viggiani P (1995) Caratteristiche fenologiche di alcune piante infestanti di recente diffusione in Italia. Informatore Fitopatologico 4: 18–23

Warwick SI (1990) Allozyme and life history variation in five northwardly colonizing North American weed species. Plant Systematic and Evolution 169: 41–54

Warwick SI and Black LD (1986) Genecological variation in recently established populations of *Abutilon theophrasti* (velvetleaf). Canadian Journal of Botany 64: 1632–1643

Warwick SI and Black LD (1988) The biology of Canadian Weeds. 90. *Abutilon theophrasti.* Canadian Journal Plant Sciences 68: 1069–1085

Zanin G and Sattin M (1988) Threshold level and seed production of velvetleaf (*Abutilon theophrasti* Medicus) in maize. Weed Research 28: 347–352

Zaragoza C (1982) Dinámica de la flora adventicia sometida al uso de herbicidas. Proceedings of the VIII Jornadas de Productos Fitosanitarios del Instituto Químico de Sarriá: 1–9, Barcelona

Biological Invasions (2005) 7: 29–35

Short-term responses to salinity of an invasive cordgrass

Jesús M. Castillo[1], Alfredo E. Rubio-Casal[1], Susana Redondo[1], Antonio A. Álvarez-López[1], Teresa Luque[1], Carlos Luque[2], Francisco J. Nieva[2], Eloy M. Castellanos[2,*]& Manuel E. Figueroa[1]
[1]*Departamento de Biología Vegetal y Ecología, Universidad de Sevilla, Apartado 1095, Spain;* [2]*Departamento de Ciencias Agroforestales, Universidad de Huelva, Spain; *Author for correspondence (e-mail: manucas@us.es; fax: +34-954-615-780)*

Received 4 June 2003; accepted in revised form 30 March 2004

Key words: chlorophyll fluorescence, gas exchange, Gulf of Cádiz, leaf expansion, salt marsh, South American neophyte, *Spartina densiflora*, water potential

Abstract

Salinity is one of the main chemical factors in salt marshes. Studies focused on the analysis of salinity tolerance of salt marsh plants are very important, since they may help to relate their physiological tolerances with distribution limits in the field. *Spartina densiflora* is a South America cordgrass, which has started its invasion of the European coastline from the southwestern Iberian Peninsula. In this work, short-term responses in adult tussocks of *S. densiflora* from southwestern Spain are studied over a wide range of salinity in a greenhouse experiment. Our results point out that *S. densiflora* has a high tolerance to salinity, showing high growth and net photosynthesis rates from 0.5 to 20 ppt. *S. densiflora* showed at the lowest salinity (0.5 ppt) high levels of photoinhibition, compensated by higher levels of energy transmission between photosystems. Adaptative mechanisms, as those described previously, would allow it to live in fresh water environments. At the highest salinity (40 ppt), *S. densiflora* showed a high stress level, reflected in significant decreases in growth, net photosynthesis rate and photochemical efficiency of Photosystem II. These responses support *S. densiflora* invasion patterns in European estuaries, with low expansion rates along the coastline and faster colonization of brackish marshes and river banks.

Abbreviations: A – net photosynthesis rate; Chl – chlorophyll; F_0 – basal fluorescence; F_p – peak of fluorescence; F_v – variable fluorescence; F_v/F_p – potential photochemical efficiency of PSII; G_s – stomatal conductivity rate; PPFD – photosynthetic photon flux density; PS II – Photosystem 2; $T_{1/2}$ – half-time for transition from F_0 to F_p; Ψ_{leaf} – leaf water potential

Introduction

Coastal ecosystems, such as salt marshes, are one of the areas most affected by the introduction of alien species. The genus *Spartina* counts different species that behave as salt marsh invaders all around the world. Their invasion patterns seem to depend on complex relationships between biological interactions with autochthonous species and habitat physical conditions such as salinity (Kittelson and Boyd 1997; Hacker et al. 2001). Thus, salinity is one of the main chemical factors in salt marshes, determining vegetation distribution with respect to elevation (Banerjee 1993) and distance to the sea (Wilson et al. 1996), through species tolerance to ion concentration and modulation of the outcomes of interspecific interactions (Broome et al. 1995; Gough and

Grace 1998). In this context, works focused on the analysis of salinity tolerance of salt marsh invaders are very important since they may help to relate physiological tolerances with distribution limits in the field (Rozema et al. 1988).

Spartina densiflora is such a species, which is invading Europe (Castillo et al. 2000), northwest Africa (Fennane and Mathez 1988) and North America (Kittelson and Boyd 1997) from South American marshes (Mobberley 1956). In Europe, its invasion started in the Gulf of Cádiz (southwestern Iberian Peninsula), where it has colonized at least eight estuaries and very contrasted habitats, from low to high marshes. However, studies on the tolerance of this species to environmental stress are very scarce (Kittelson and Boyd 1997; Nieva et al. 1999; Castillo et al. 2000).

The work described in this paper aimed to analyze short-term performance of *S. densiflora* from the southwestern Iberian Peninsula over a wide range of salinity. The specific objectives were to examine water potential, growth, leaf gas exchange and chlorophyll fluorescence in adult *S. densiflora* tussocks in five salinity treatments, from fresh water to hypersalinity, in a greenhouse experiment. We also sought to compare whether these measures of performance were in accordance with its actual distribution and discuss possible evolution of its invasion depending on its salinity tolerance. Our results may help to understand invasion mechanisms of cordgrasses.

Growth, leaf gas exchange and chlorophyll fluorescence have been identified as adequate tools to analyze stress due to salinity (Mishra et al. 1991; Belkhodja et al. 1994; Ewing et al. 1997).

Materials and methods

General methods

Spartina densiflora adult tussocks were collected, in June 1998, in Odiel river salt marshes (southwestern Iberian Peninsula) from 'Punta del Sebo' low marsh. Plants were transplanted into clean sand culture in small plastic pots for propagation in a greenhouse (Ewing et al. 1995). For the experiment, plants (5–18 live stems per tussock) were repotted in expands perlite (Floreal. Agroperlita F-3) in 11 cm diameter pots (Trovadec plastics).

Plants were grown using five salinity treatments (0.5, 10, 15, 20 and 40 ppt), immersing the pots 2 cm in a saline/nutrient solution (20% modified Hoagland's solution). The salinity treatments were obtained by using Hoagland's Solution, and Hoagland's Solution plus sea salts (Instant Ocean) to give 10, 15, 20 and 40 ppt. The last treatment was obtained progressively, though the treatment of 20 ppt took 1 week. Concentrations of the full-strength nutrient solution were 1.02 ppt KNO_3, 0.49 ppt $Ca(NO_3)_2$, 0.23 ppt $NH_4H_2PO_4$, 0.49 ppt $MgSO_4 \cdot 7H_2O$, 2.86 ppt H_3BO_3, 1.81 ppt $MnCl_2 \cdot 4H_2O$, 0.08 ppt $CuSO_4 \cdot 5H_2O$, 0.22 ppt $ZnSO_4 \cdot 7H_2O$, 0.09 ppt $H_2MoO_4 \cdot H_2O$, 0.6 ml $FeSO_4 \cdot 7H_2O$ 0.5% and 0.6 ml tartaric acid 0.4%. The solution was changed once a week. Five replicates of each treatment were used. The experiment was carried out for 29 days in September 2000, in a greenhouse with controlled temperatures between 21 and 25 °C; the photoperiod was extended to 15 h with the use of incandescent lights (Osram Vialox NAV-T (SON-T) 400 W) with a photosynthetically active photon flux density (PPFD) of 250 $\mu E/m^2$ s at canopy level.

The following measurements were made: leaf water potential, leaf expansion, leaf gas exchange and chlorophyll *a* fast kinetic fluorescence parameters.

Leaf water potential

Leaf water potential (Ψ_{leaf}) was measured by a Scholander bomb during the noon solar hour on adult leaves of each of four tillers (each from a different clump, chosen at random) in every treatment, at the end of the experiment.

Leaf expansion

Leaf expansion was measured on five tillers (each from a different clump, chosen at random) in every treatment at the end of the experiment by placing a marker of inert sealant on the base of the youngest accessible leaf. The distance between the marker and the leaf base was measured at the beginning and the end of a 3-day period (Ewing et al. 1995).

Gas exchange

Measurements of the net photosynthesis rate (A) and stomatal conductivity rate (G_s) were carried out at the end of the experiment on the second youngest leaf of each of three tillers (each from a different clump, chosen at random) in every treatment after 2 h of lighting with a PPFD of 250 μE/m^2 s at canopy level. Measurements were made using a portable infrared CO_2 analyzer (ADC LCA-3, Analytical Development Co. Ltd, Hoddesdon, UK) in differential mode and in an open circuit; it was coupled to a Parkinson Leaf Chamber (ADC PLC-3N), using a radiant flux density of 210 μE/m^2 s (Long and Hällgren 1993).

Chlorophyll fluorescence

Chlorophyll *a* fast kinetic fluorescence was measured on the second-youngest leaf of each of five tillers (each from a different clump, chosen at random) at every treatment, in the prevailing air temperature at PPFD of 250 μE/m^2s. Chlorophyll fluorescence measurements were made with a portable non-modulated fluorimeter (Plant Stress Meter, PSM Mark II, Biomonitor S.C.I. AB, Umeå, Sweden) and white leaf enclosures (Biomonitor 1020) for dark adaptation. Details of the instrument are provided by Öquist and Wass (1988). Leaves were dark-adapted for 30 min before measurements of the fluorescence transient over 2 s and with an actinic stimulation at PPFD of 400 μE/m^2 s.

The initial fluorescence (F_0) which depends on the size of the PSII chlorophyll antenna and on the functional integrity of PSII reaction centers (Krause and Weis 1991) was determined by the shutter aperture, which is fast enough to give satisfactory resolution. A flash of actinic light yielded a peak of fluorescence (F_p) for this light level. The half-time for transition from F_0 to F_p ($T_{1/2}$) was determined, and it is related to reduction rate of Q_A, Q_B and PQ; in fact, it has been used to determine the amount of functional PSII centers and the size of the PQ pool (Bolhàr-Nordenkampf and Öquist 1993). The ratio of variable to peak fluorescence ($F_v/F_p = (F_p-F_0)/F_p$) was used as a measure of the maximum photochemical yield of PSII; this ratio correlates with the number of functional PSII reaction centers and was used to quantify photoinhibition (Krivosheeva et al. 1996).

Data analysis

Analysis was carried out using 'Statistica' release 5.1 (Statsoft Inc.). Pearson correlation coefficients were calculated between salinities and physiological variables. Physiological measurements were compared between treatments by a one-way analysis of variance. The least significant differences (LSD) between means were calculated only if the F-test was significant at the 0.05 level of probability. Data were tested for homogeneity of variance with the Levene test ($P > 0.05$). Deviations were calculated as the standard error of the mean (SEM).

Results

Leaf water potential and leaf expansion

Ψ_{leaf} was very highly negatively correlated with salinity concentration ($r = -0.96$, $P = 0.01$, $n = 5$), oscillating between −0.52 MPa at the lowest salinity (0.5 ppt) and −3.47 MPa at the highest one (40 ppt) (Figure 1).

Leaf expansion was as well negatively correlated with salinity concentration ($r = -0.92$, $P = 0.03$, $n = 5$). The highest leaf expansion was recorded at the lowest salinity (2.55 ± 0.14 cm/day), which was similar to those recorded at 10, 15 and 20 ppt and significantly higher than that

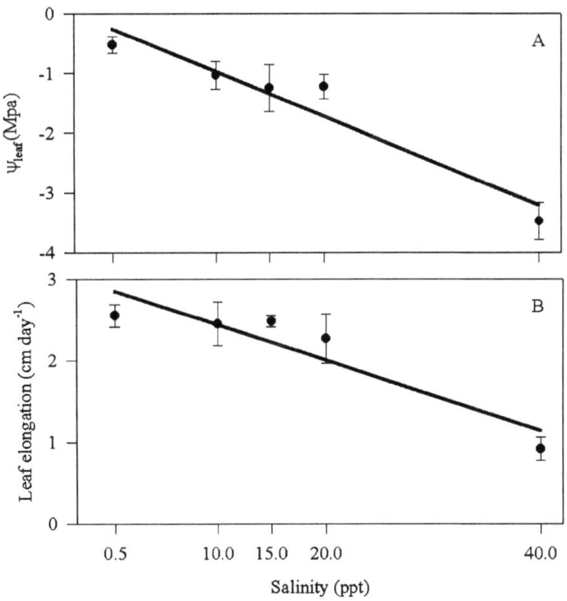

Figure 1. Leaf water potential (Ψ_{leaf}) (A) and leaf elongation (B) in relation to salinity in *S. densiflora* from the southwestern Iberian Peninsula. Regression equations ($n = 5$): Ψ_{leaf}, $y = -0.22 - 0.07x$; leaf elongation, $y = 2.87 - 0.04x$.

recorded at 40 ppt (LSD post-hoc test, $P = 0.001$; ANOVA, $F = 11.4$, $P < 0.001$) (Figure 1).

Gas exchange

G_s oscillated between 95.6 ± 16.0 and 148.6 ± 21.3 mmol/m^2 s, without significant differences between treatments ($F = 1.53$, $P < 0.266$) (Figure 2).

The net photosynthesis rate was negatively correlated with salinity ($r = -0.98$, $P = 0.01$, $n = 5$), decreasing to half from fresh water (16.6 ± 0.7 µmol/m^2 s) to hypersalinity (7.6 ± 1.0 µmol/m^2 s) (Figure 2). On the other hand, A was positively correlated with Ψ_{leaf} ($r^2 = 0.99$, $P = 0.01$, $n = 5$) and leaf expansion ($r^2 = 0.96$, $P = 0.01$, $n = 5$).

Chlorophyll fluorescence

The ratio F_v/F_p showed a minimum at the lowest and the highest salinity concentrations (ANOVA, $F = 3.9$, $P < 0.017$). $T_{1/2}$ was significantly higher at 0.5, 10 and 40 ppt than at 15 and 20 ppt (ANOVA, $F = 5.8$; $P < 0.003$), oscillating between 243 and 294 ms. The initial fluorescence (F_0) did not vary significantly with salinity, oscillating around 0.05 in every treatment. F_p and F_v showed minimum values at the lowest and the highest salinities; however, these differences were not significant with respect to the other treatments (ANOVA, $F = 2.1$, $P < 0.114$; $F = 2.3$, $P < 0.097$; respectively) (Figure 3).

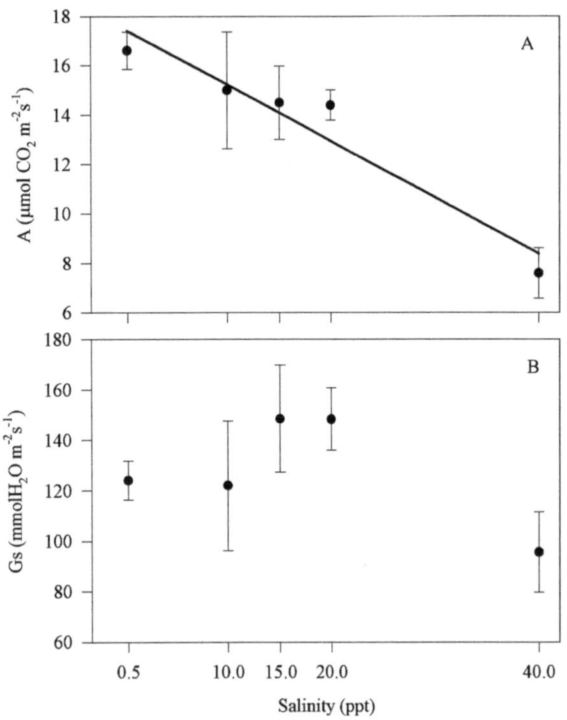

Figure 2. (A) Net photosynthesis rate (A) and (B) stomatal conductance (G_s) in relation to salinity in *S. densiflora* from SW Iberian Peninsula. Regression equation (*n = 5*): A, $y = 17.45 - 0.22x$.

Discussion

This study shows that adult tussocks of the invasive cordgrass, *Spartina densiflora*, from the southwestern Iberian Peninsula show a high short-term tolerance to salinity with high values in growth and net photosynthesis rate from 0.5 to 20 ppt. These results are in agreement with *S. densiflora* distribution in invaded European estuaries, where it colonizes habitats with contrasted salinity regimens as low and brackish marshes and saltpans (Nieva et al. 2001). The high acclimation capacity of *S. densiflora* to salinity was also reflected in the regulation of its Ψ_{leaf}, behaving as an osmo regulator (Nieva et al. 1999).

S. densiflora showed a lower F_v/F_p ratio at the lowest salinity than at intermediate salinities, which resulted from a decrease in F_p. This photo inhibition in fresh water would be due to the dissipation processes of excessive energy, related to non-photochemical quenching such as thermal deactivation and xanthophyll cycle (Griffiths and Maxwell 1999), and/or a limitation in Chl synthesis. It has been described that NaCl increases the Chl content in coastal species (Beer et al. 1976). This photoinhibition would be compensated by adaptative biochemical mechanisms, such as higher levels of energy transmission between photosystems through maintenance of an abundant and active PQ pool – higher $T_{1/2}$ values – (Fernández-Baco et al. 1998). These mechanisms would allow *S. densiflora* to keep high net photosynthesis and growth rates in fresh and brackish water environments.

At the highest salinity (40 ppt), a high stress level was recorded in *S. densiflora*, reflected in significant decreases in growth, net photosynthesis rate and photochemical efficiency of PS II. This decrease in A was high (around 50%

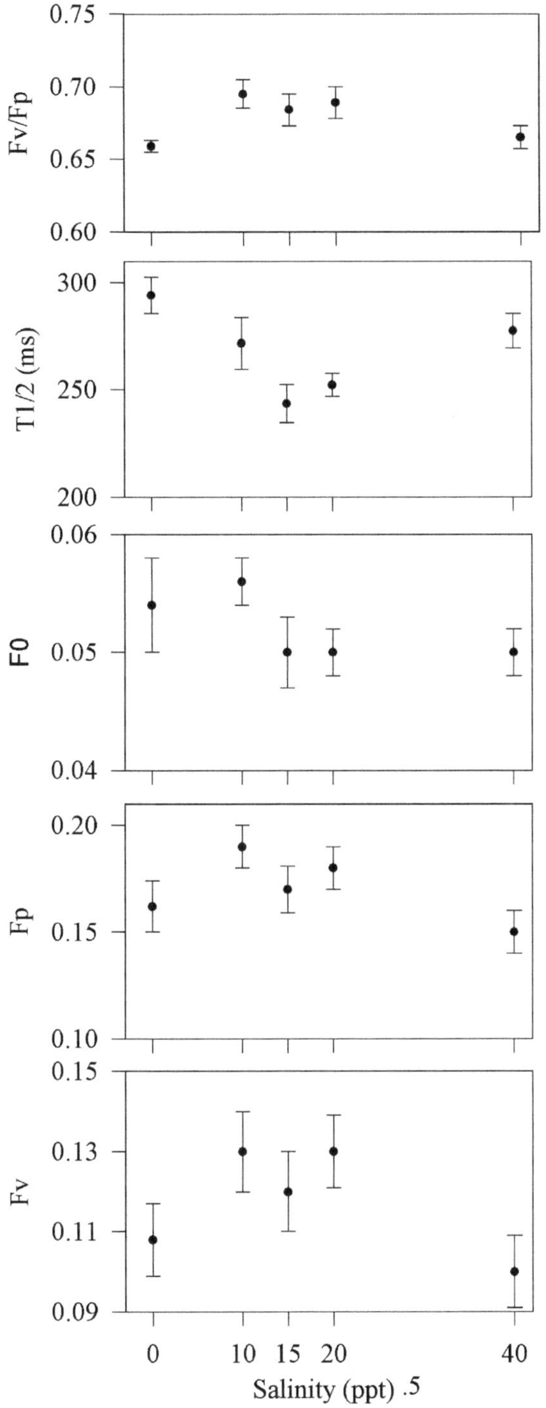

Figure 3. Potential photochemical efficiency of PSII (F_v/F_p), half-time for transition from F_0 to F_p ($T_{1/2}$), initial fluorescence (F_0), peak fluorescence (F_p) and variable fluorescence (F_v) in relation to salinity in *S. densiflora* from the southwestern Iberian Peninsula.

compared with fresh water treatment) as it used to be in C_4 species (Drake 1989) and it was independent of G_s values. It has been described as the primary effect of salinity on metabolic events (Drake 1989) through an excess of ions in the plant tissue and/or an inhibition of nutrient ion uptake (Greenway and Munns 1980). These alterations interfere with metabolic functions such as protein synthesis, which determinates photoinhibition levels (Ohad et al. 1985), and they lead to biochemical changes that affect carboxylase activity of the ribulose-1,5-bisphosphate carboxylase/oxygenase (Antolín and Sánchez-Díaz 1993). Salt stress was also reflected in a significant decrease of F_v/F_p. The nature of this photoinhibition seemed to be related with photoprotective mechanisms such as xanthophyll cycle, since a slight decrease in F_p and F_v was recorded. As well as in fresh water, *S. densiflora* photosynthetic apparatus showed adaptation mechanisms to salinity stress reflected in a significant increase of $T_{1/2}$ at the highest salinity, suggesting high levels of energy transmission between photosystems (Fernández-Baco et al. 1998). This adaptation capacity would not be enough to compensate salt stress effects, since, as pointed out previously, other processes such as the Calvin cycle may also be involved.

Our results support *S. densiflora* invasion patterns in European estuaries, with low expansion rates near the sea, and an active colonization of brackish marshes and river banks, where it competes with native oligohaline species (Nieva 1996). Similar results were found by Kittelson and Boid (1997) working in adult *S. densiflora* tussocks in invaded Californian marshes. Nieva et al. (1999), working with invading *S. densiflora* in the Gulf of Cádiz, found an increase in A with a decrease in salinity, and Valiela et al. (1978) suggested that salinity might be limiting the growth of *Spartina alterniflora* wild populations. All these results suggest that *S. densiflora* invasion would be favored in brackish environments and by anthropogenic disturbances lowering salinity, such as fresh water emissions.

The elevated salinity tolerance of *S. densiflora* together with its high competitive ability – growing in very dense tussocks with a phalanx growth (Figueroa and Castellanos 1988) – would allow this neophyte to invade a wide range of habitats displacing native species, such as other invasive weeds in salt marshes (Rea and Storrs 1999; Merriam and Feil 2002). Studies rooted to analyze competitive interactions of *S. densiflora* with native species in relation with environmental conditions are needed to clarify how far this alien species is affecting autochthonous communities.

Acknowledgements

This study was carried out through an investigation agreement with the Port Authority of Huelva and the General Directorate of Environmental Protection of the Junta de Andalucía. We also thank the DIGICYT for grant PB-1455, the Directorate of the Odiel Marshes Natural Park, the Andalusian Research Plan and Seville University Greenhouse General Service for collaboration. We wish to thank Dr Lara very much for her comments on physiology.

References

Antolín MC and Sánchez-Díaz M (1993) Effects of temporary drought on photosynthesis of alfalfa plants. Journal of Experimental Botany 44: 1341–1349

Banerjee LK (1993) Influence of salinity on mangrove zonation. In: Lieth H and Al Masoom A (eds) Towards the Rational Use of High Salinity Tolerant Plants, Vol 1, pp 181–186. Kluwer Academic Publishers, Dordrecht, The Netherlands

Beer S, Shomer A and Waisel Y (1976) Salt stimulated phosphoenol pyruvate carboxylase in *Cakile maritima*. Physiologia Plantarum 34: 293–295

Belkhodja R, Morales F, Abadía A, Gómez-Aparisi J and Abadía J (1994) Chlorophyll fluorescence as a possible tool for salinity tolerance screening in Barley (*Hordeum vulgare* L.). Plant Physiology 104: 667–673

Bolhàr-Nordenkampf HR and Öquist G (1993) Chlorophyll fluorescence as a tool in photosynthesis research. In: Hall DO, Scurlock JMO, Bolhàr-Nordenkampf HR, Leegoog RC and Long SP (eds) Photosynthesis and Production in a Changing Environment: a Field and Laboratory Manual, pp 193–206. Chapman & Hall, London

Broome SW, Mendelssohn IA and McKee KL (1995) Relative growth of *Spartina patens* (Ait.) Muhl. and *Scirpus olneyi* gray occurring in a mixed stand as affected by salinity and flooding depth. Wetlands 15: 20–30

Castillo JM, Fernández-Baco L, Castellanos EM, Luque CJ, Figueroa ME and Davy AJ (2000) Lower limits of *Spartina densiflora* and *S. maritima* in a Mediterranean salt marsh determinated by different ecophysiological tolerances. Journal of Ecology 88: 801–812

Drake BG (1989) Photosynthesis of salt marsh species. Aquatic Botany 34: 167–180

Ewing K, McKee K, Mendelssohn I and Hester M (1995) A comparision of indicators of sublethal salinity stress in the salt marsh grass, *Spartina patens* (Ait.) Muhl. Aquatic Botany 52: 59–74

Ewing K, McKee K and Mendelssohn I (1997) A field comparision of indications of sublethal stress in the salt-marsh grass *Spartina patens*. Estuaries 20: 48–65

Fennane M and Mathez J (1988) Nouveaux materiaux pour la flore du Maroc. Naturalia Monspeliensia 52: 135–141

Fernández-Baco L, Figueroa ME, Luque C and Davy AJ (1998) Diurnal and seasonal variations in chlorophyll fluorescence in two Mediterranean-grassland species under field conditions. Photosynthetica 35: 535–544

Figueroa ME and Castellanos EM (1988) Vertical structure of *Spartina maritima* and *Spartina densiflora* in Mediterranean marshes. In: Werger MJA, van der Aart PJM, During HJ and Verhoeven JTA (eds) Plant Form and Vegetation Structure, pp 105–108. SPB Academic Publishing, The Hague, The Netherlands

Greenway H, Munns R (1980) Mechanisms of salt tolerance in non-halophytes. Annual Review of Plant Physiology 31: 149–190

Griffiths H and Maxwell K (1999) In memory of C.S. Pittendrigh. Does exposure in forest canopies relate to photoprotective strategies in epiphytic bromeliads? Functional Ecology 13: 15–23

Gough L and Grace JB (1998) Effects of flooding, salinity and herbivory on coastal plant communities, Louisiana, United States. Oecolgia 117: 527–535

Hacker SD, Heimer D, Hellquist CE, Reeder TG, Reeves B, Riordan TJ and Dethier MN (2001) A marine plant (*Spartina anglica*) invades widely varying habitats: potential mechanisms of invasion and control. Biological Invasions 3: 211–217

Kittelson PM and Boyd MJ (1997) Mechanism of expansion for an introduced species of cordgrass, *Spartina densiflora*, in Humbolt Bay, California. Estuaries 20: 770–778

Krause GH and Weis E (1991) Chlorophyll fluorescence and photosynthesis: the basics. Annual Review of Plant Physiology and Plant Molecular Biology 42: 313–349

Krivosheeva A, Tao DL, Ottander C, Wingsle G, Dube SL and Öquist G (1996) Cold acclimated and photoinhibition of photosynthesis in Scots pine. Planta 200: 296–305

Long SP and Hällgren JE (1993) Measurement of CO_2 assimilation by plants in the field and the laboratory. In: Hall DO, Scurlock JMO, Bolhàr-Nordenkampf HR, Leegood RC and Long SP (eds) Photosynthesis and Production in a Changing Environment. A Field and Laboratory Manual, pp 129–167. Chapman & Hall, London

Merriam RW and Feil E (2002) The potential impact of an introduced shrub on native plant diversity and forest regeneration. Biological Invasions 4: 369–373

Mishra SK, Subrahmanyam D and Singhal GS (1991) Interrelationship between salt and light stress on primary processes of photosynthesis. Journal of Plant Physiology 138: 92–96

Mobberley DG (1956) Taxonomy and distribution of the genus *Spartina*. Iowa State College Journal of Science 30: 471–574

Nieva FJJ (1996) Aspectos ecológicos de *Spartina densiflora* Brong. Master's Thesis, University of Seville, Seville, Spain, 241 pp

Nieva FJJ, Castellanos EM, Figueroa ME and Gil F (1999) Gas exchange and chlorophyll fluorescence of C_3 and C_4 saltmarsh species. Photosynthetica 36: 397–406

Nieva FJ, Diaz-Espejo A, Castellanos EM and Figueroa ME (2001) Field variability of invading populations of *Spartina densiflora* Brong grown in different habitats of the Odiel marshes (SW Spain). Estuarine, Coastal and Shelf Science 52: 515–527

Ohad I, Kyle DJ and Hirschberg J (1985) Photorespiration and photoinhibition. Some implications for the energetics of photosynthesis. Biochemical and Biophysical Acta 639: 77–98

Öquist G and Wass R (1988) A portable microprocessor operated instrument for measuring chlorophyll fluorescence kinetics in stress physiology. Plant Physiology 73: 211–217

Rea N and Storrs MJ (1999) Weed invasions in wetlands of Australia's Top End: reasons and solutions. Wetlands Ecology and Management 7: 47–62

Rozema J, Scholten MCT, Blaauw PA and van Diggelen J (1988) Distribution limits and physiological tolerances with particular reference to the salt marsh environment. In: Davy AJ, Hutchings MJ and Watkinson AR (eds) Plant Population Ecology, pp 137–164. Blackwell, Oxford

Valiela I, Teal JM and Deuser WG (1978) The nature of growth forms in the salt marsh grass *Spartina alterniflora*. American Naturalist 112: 461–470

Wilson JB, King W McG, Sykes MT and Partridge TR (1996) Vegetation zonation as related to the salt tolerance of species of brackish riverbanks. Canadian Journal of Botany 74: 1079–1085

Biological Invasions (2005) 7: 37–47

The American brine shrimp as an exotic invasive species in the western Mediterranean

Francisco Amat[1,*], Francisco Hontoria[1], Olga Ruiz[1], Andy J. Green[2], Marta I. Sánchez[2], Jordi Figuerola[2] & Francisco Hortas[3]
[1]*Instituto de Acuicultura de Torre de la Sal (CSIC), 12595 Ribera de Cabanes (Castellón), Spain;* [2]*Estación Biológica de Doñana (CSIC), Avda. María Luisa s/n, Pabellón del Perú, 41013 Sevilla, Spain;* [3]*Grupo de Conservación de Humedales Costeros, Departamento de Biología, Facultad de Ciencias del Mar y Ambientales, Apartado 40, 11510 Puerto Real (Cádiz), Spain; *Author for correspondence (e-mail: amat@iats.csic.es; fax: +34-964-319509)*

Received 4 June 2003; accepted in revised form 30 March 2004

Key words: *Artemia*, aquaculture, salterns, western Mediterranean

Abstract

The hypersaline environments and salterns present in the western Mediterranean region (including Italy, southern France, the Iberian Peninsula and Morocco) contain autochthonous forms of the brine shrimp *Artemia*, with parthenogenetic diploid and tetraploid strains coexisting with the bisexual species *A. salina*. Introduced populations of the American brine shrimp *A. franciscana* have also been recorded in these Mediterranean environments since the 1980s. Based on brine shrimp cyst samples collected in these countries from 1980 until 2002, we were able to establish the present distribution of autochthonous brine shrimps and of *A. franciscana*, which is shown to be an expanding invasive species. The results obtained show that *A. franciscana* is now the dominant *Artemia* species in Portuguese salterns, along the French Mediterranean coast and in Cadiz bay (Spain). Co-occurrence of autochthonous (parthenogenetic) and American brine shrimp populations was observed in Morocco (Mar Chica) and France (Aigues Mortes), whereas *A. franciscana* was not found in Italian cyst samples. The results suggest these exotic *A. franciscana* populations originate as intentional or non-intentional inoculations through aquacultural (hatchery effluents) or pet market activities, and suggest that the native species can be rapidly replaced by the exotic species.

Introduction

The brine shrimp *Artemia* (Branchiopoda, Anostraca) is perhaps the most conspicuous inhabitant of hypersaline lakes and lagoons and solar saltern ponds, coastal and inland, over the world, where simple trophic structures and low species diversity are present (Lenz and Browne 1991).

The genus *Artemia* comprises a group of bisexual and parthenogenetic species, which probably diverged five to six million years ago from an ancestral form living in the Mediterranean area (Abreu-Grobois and Beardmore 1982; Badaracco et al. 1987). The string of shallow briny lakes into which the Mediterranean sea had converted (Hsü et al. 1977) created opportunities for colonization, extinction and recolonization cycles with different degrees of reproductive isolation, while the appearance of a parthenogenetic mode of reproduction, together with polyploidy, may have facilitated dispersion. Thus, the Mediterranean has been proposed as the centre of radiation for *Artemia*, based on changes in reproduction modes, bisexuality and parthenogenesis on the one hand, together with diploidy and polyploidy on the other (Gajardo et al. 2002).

On the basis of criteria from morphometry and laboratory reproductive isolation, and subsequently through karyology, allozyme divergence and new molecular (DNA) markers, seven bisexual species and two or three parthenogenetic forms are currently recognized in the genus *Artemia*. They all look rather similar in body shape, but show morphological traits that enable morphometric differentiation when they are cultured under standard laboratory conditions (Hontoria and Amat 1992a, b).

The bisexual *A. persimilis* (Piccinelli and Prosdocimi 1968) (Argentina and Chile) and *A. franciscana* (Kellogg 1906) (North, Central and South America) are endemic to the New World. The bisexual *A. salina* Leach 1819 (Mediterranean area and Africa), *A. urmiana* (Günther 1890) from Iran, *A. sinica* (Cai 1989) from P.R. China, and *A. tibetiana* (Abatzopoulos et al. 1998, 2002) from Tibet, with *Artemia* sp. from Kazakhstan (Pilla and Beardmore 1994) are endemic to the Old World. Recently, the American species *A. franciscana* has been introduced in the Old World, especially in the Far East and in the Mediterranean area, as explained in this study.

The parthenogenetic forms, with different degrees of ploidy, are present in the Old World, i.e. Eurasia and Africa, and were introduced in Australia. Although these forms are listed taxonomically with the binomen *Artemia parthenogenetica*, the wide diversity found among different asexual populations, especially in terms of ploidy, suggests that their grouping under a single species may be misleading (Browne et al. 1991; Gajardo et al. 2002). The distribution of *Artemia* populations in the western Old World, including Italy, south of France and the Iberian Peninsula (Spain and Portugal), together with the north of Africa, is especially interesting owing to the presence and distribution of the Mediterranean bisexual *A. salina* and, at least, two different parthenogenetic forms, diploid and tetraploid (Artom 1906; Stella 1933; Gilchrist 1960; Stefani 1960; Amat 1983a, b ; Vieira and Amat 1985; Vanhaecke et al. 1987; Amat et al. 1995; Tryantaphyllidis et al. 1997a, b).

This region also shows the unfortunate event of the presence of the American species *Artemia franciscana* (Narciso 1989; Hontoria et al. 1987; Amat et al. 1995). This paper aims to review the current distribution of *A. franciscana* populations in the western Mediterranean region and their likely origins. We also compare the biometry of both introduced and autochthonous populations, and consider the impact of the exotic species on the native ones.

Materials and methods

This research was carried out using a large collection of *Artemia* cyst samples in the Instituto de Acuicultura de Torre de la Sal (CSIC), supported by the samples available from the *Artemia* Reference Center (University of Ghent, Belgium), and a database of morphometric characterizations of adult specimens from different *Artemia* populations and species. This database used a 'morphometric standard' describing the populations obtained after hatching these cysts in the laboratory and their culture under standard conditions (Hontoria and Amat 1992a).

The cyst collection contained about 130 cyst samples from Western Europe (Spain, Portugal and Italy) collected since 1980. During 2001 and 2002, new cyst samples were taken from the southwest of Spain and Portugal (by the authors), and the southeast of France (provided by Dr Thomas Lenormand, Université de Montpellier, France), the European regions where the invasive presence of *A. franciscana* was first reported (Narciso 1989; Thiery and Robert 1992). We also analysed cysts of asexual *Artemia* that we extracted from the pellets of Redshank *Tringa totanus* that were collected from Cadiz Bay in July and August 2002 (A.J. Green, M.I. Sánchez, F. Amat, J. Figuerola, F. Hontoria, O. Ruiz, F. Hortas, unpublished manuscript).

Cysts from Italy (Dr Graziella Mura, University of Rome) and from Portugal (Dr Mª Elena Vilela, Instituto de Investigaçao das Pescas e do Mar, Lisbon) were preserved in plastic bags under vacuum. Other cysts samples were processed according to standardized methods (Vanhaecke and Sorgeloos 1980) and stored at 4 °C in sealed plastic bags.

Biometry of adults

The nauplii obtained by cyst hatching were made to grow up in 5 l plastic containers, with 70 g l^{-1} filtered brine (seawater plus crude sea salt), and put on a mixed diet of *Dunaliella salina* and *Tetraselmis suecica*. The temperature was maintained at 24 ± 1 °C, under mild aeration at a 12D : 12L photoperiod. The medium was monitored and renewed every 2 days. Once 50% of the females attained full ovisac development and the first ovoviviparous offspring was observed, random samples of 30 females (parthenogenetic and bisexual strains) and 30 males (bisexual strains) were removed from the culture, anaesthetized and measured under a dissecting microscope. The following morphological parameters were quantified in each specimen: total length, abdominal length, width of third abdominal segment, width of the ovisac in females and width of the genital segment in males, length of furca, number of setae inserted on each branch of the furca, width of head, maximal diameter and distance between compound eyes, length of first antenna and the ratio of abdominal length × 100/ total length. The biometrical analysis of these data was performed via multivariate discriminant analysis (Hontoria and Amat 1992a) using the statistical package SPSS for Windows version 11.0 (SPSS Inc., Chicago, Illinois, USA), and the results were integrated in the morphological data base developed at the Instituto de Acuicultura de Torre de la Sal (Amat et al. 1995).

Biometry of cysts

When cysts from old or mishandled samples did not hatch, it was impossible to obtain living nauplii; therefore, to grow up laboratory populations to adulthood. In this case, the population specific adscription was obtained through the biometric study of cysts, which provides a fitting alternative for this purpose (Vanhaecke and Sorgeloos 1980; Hontoria 1990).

Samples providing sufficient quantities of cysts allowed cyst diameter analysis with an electronic Coulter Counter® counter-sizer (Vanhaecke and Sorgeloos 1980; Hontoria 1990). The other cyst samples were previously hydrated with 20 g l^{-1} filtered seawater, at 28 °C with continuous illumination and aeration, and measured under a dissecting microscope to the nearest μm. Cysts were hatched under standard conditions: 35 g l^{-1} filtered sea water, at 28 °C, with continuous illumination and aeration.

Several cyst samples did not hatch after the first attempt, and according to the amount of cysts available, they were submitted to two or three hydration and dehydration (24 h oven dessication under 39 °C) cycles, and/or to a H_2O_2 treatment, processes that terminate diapause of *Artemia* cysts (Lavens and Sorgeloos 1987; VanStappen et al. 1998), before a final attempt to hatch them in order to obtain living nauplii.

Results

Biometry of cysts and adults

The information obtained from cyst samples from Portugal is shown in Table 1. Only those samples collected since 1990 hatched. Laboratory populations showed the exclusive presence of *A. franciscana*, based on the morphometric study of adult specimens (Figure 1). However, the information from cyst biometry allows inferring the presence of *A. franciscana* in the Algarve from the beginning of the 1980s. In the Sado estuary area, the situation is similar, but Olhos and Cachopos salterns may still have held *A. parthenogenetica* populations in the 1980s according to the cyst diameter that exceeded 260 μm (Hontoria 1990). In the Tejo estuary, it is possible to infer the presence of *A. franciscana* in Alcochete and Boavista salterns in the 1980s, but autochthonous parthenogenetic (diploid and tetraploid) populations (cyst diameter between 260 and 280 μm) were dominant at that time. Finally, in the Esmolas salterns, from the district of Aveiro, the presence of *A. parthenogenetica* was stated by Vieira (1990), but samples collected in 1991 showed the exclusive presence of *A. franciscana*.

More recent cyst samples from Huelva and Cadiz provinces in Spain, from Mar Chica (Nador, Morocco) and from the South of France hatched successfully. The populations obtained from these cysts verified the presence of the autochthonous bisexual (*A. salina*) and the parthenogenetic diploid and tetraploid strains, together with the exotic *A. franciscana* (Table 2). The Westernmost Spanish populations (Odiel in Huelva and N.S. del Rocío in Sanlucar de Barrameda, Cadiz) showed the exclusive presence of autochthonous populations (*A. salina* and/or *A. parthenogenetica*) in variable ratios, whereas the Moroccan

Table 1. Artemia cyst samples available from Portugal, mean diameter of cysts and taxonomical identification.

Locality	Sampling date	Cyst diameter (μm)			Observations
		Micro	C.C.	C.C.(*)	
Algarve Province					
San Francisco salterns	1980–1981	243	236	238	NH
Marina Bias salterns	1987	246	248	–	NH
Olhao salterns	1985	247	–	259	NH
Tavira salterns	1985	240	–	–	NH
Olhao salterns	2002	253	–	–	*A. franciscana*
Faro. Ludo salterns	2002	245	–	–	*A. franciscana*
Castro Marim salterns	2002	250	–	–	*A. franciscana*
Sado Estuary					
Batalha salterns	1986	236	–	–	NH
Sado salterns	1987	248	–	–	NH
Olhos salterns	1986	262	–	–	NH
Cachopos salterns	1987	276	–	–	NH
Rio Frio salterns	1993	228	–	–	*A. franciscana*
Bonfim salterns	1996	224	–	–	*A. franciscana*
Tejo Estuary					
Alcochete salterns	1988	259	256	266	NH
Boavista salterns	1987	258	260	–	NH
Marina Nova salterns	1987	264	273	–	NH
Marina Velha salterns	1987	–	276	–	NH
Providencia salterns	1987	–	276	–	NH
Aveiro District					
Esmolas salterns	1985	266 (**)	–	263	*A. parthenog.* (d)
Esmolas salterns	1991	249	–	–	*A. franciscana*
Esmolas salterns	1993	248	–	–	*A. franciscana*

Micro.: cyst diameter measured with micrometer eyepiece. C.C.: cyst diameter measured with Coulter Counter. C.C.(*): idem (Hontoria 1990); (**) Vieira 1990; NH = No hatching cysts. (d): *Artemia parthenogenetica* diploid.

population showed a mixture of the autochthonous diploid parthenogenetic and the American species. However, the salterns located in Cadiz bay showed the exclusive presence of *A. franciscana*. French samples corresponded mostly to *A. franciscana* populations, verified through specimens obtained under culture or through those provided as preserved samples. In Aigues Mortes, low proportions of the autochthonous parthenogenetic population were recorded.

Cyst samples available from Italy and the information obtained from them are shown in Table 3. Some of these samples, collected before 1987, did not hatch. Laboratory populations obtained from viable cysts showed the presence of the autochthonous populations, i.e., the bisexual *A. salina* and both parthenogenetic strains, diploid and tetraploid. According to data reported previously by Hontoria (1990), the cysts from the Comacchio salterns correspond to a parthenogenetic tetraploid strain because of their big size, while the Sicilian sample from Isola Longa (Trapani) resembles the size of cysts obtained for *A. salina* populations from Tarquinia and Sardinian salterns.

The multivariate procedure produces 12 discriminant functions for each analysis (males and females). In the case of the females, the first 11 functions significantly ($P \leq 0.01$) account for the increase of variance explained when they are included in the model. For males, the analysis needs only the first eight discriminant functions to completely separate the populations studied. These functions significantly ($P \leq 0.01$) account for the variance explained. However, in both cases, the four first discriminant functions account for the larger part of the variation (90.5% in the analysis pertaining to females and 93.2% in that for the males).

Figures 1 and 2 summarize the centroids (mean points for each population) for the first two discriminant functions obtained, for females and

FEMALES

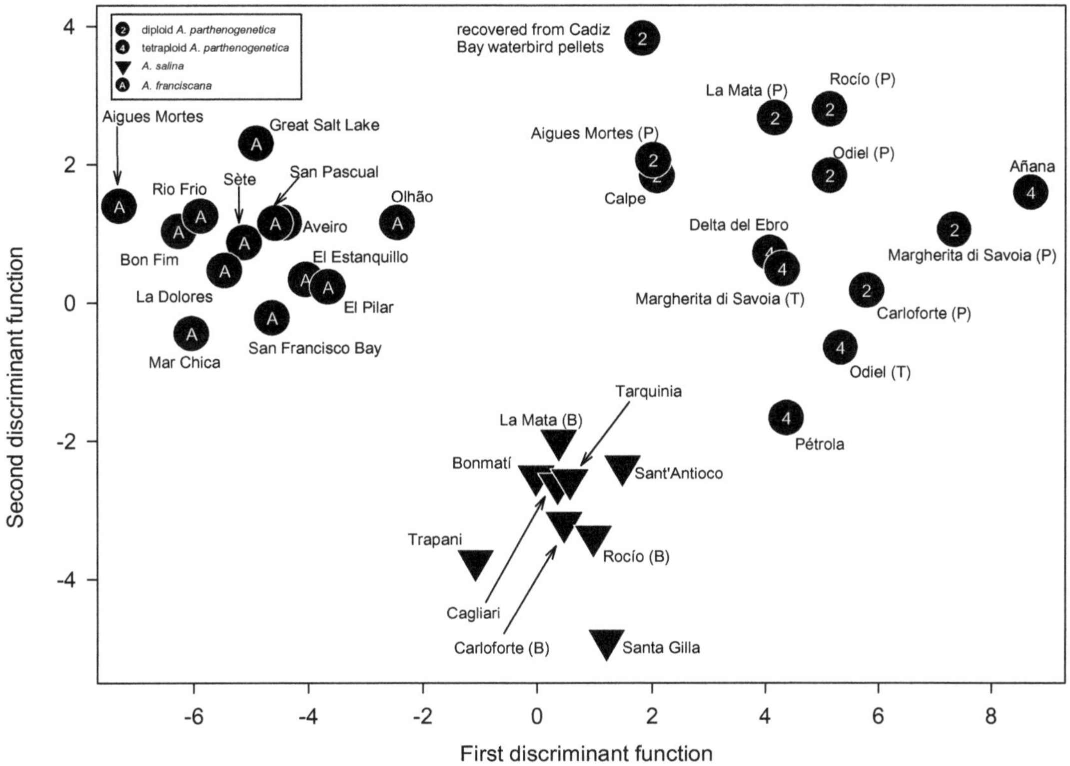

Figure 1. Group centroids of the populations studied for the first two discriminant functions resulting from the discriminant analysis on female morphometric variables.

Table 2. Artemia populations obtained in the laboratory from cyst samples collected in southwestern Spain, north of Morocco and southeastern France, and their relative specific composition where different species or strains appeared.

Locality	Sampling date	Specific composition	
Spain			
Huelva Province			
Esteros Odiel saltern	06. 2002	*A. parthenog.* (d): 97%	*A. parthenog.* (t): 3%
Cadiz Province			
N.S. del Rocío saltern	01. 2002	*A. parthenog.* (d): 98%	*A. salina* (bisex.) 2%
El Estanquillo saltern	01. 2002	*A. franciscana*	
El Pilar saltern	06. 2002	*A. franciscana*	
San Pascual saltern	02. 2003	*A. franciscana*	
La Dolores saltern	02. 2003	*A. franciscana*	
Morocco			
Laguna Mar Chica saltern	06. 2000	*A. parthenog.* (d): 80%	*A. franciscana*: 20%
France			
Sete-Listel saltern	05. 2002	*A. franciscana*	
Aigues Mortes saltern	06. 2002	*A. parthenog.* (d): 2%	*A. franciscana*: 98%
Fos saltern	05. 2002	*A. franciscana* (p. s.)	
Pesquiers saltern	05. 2002	*A. franciscana* (p. s.)	
Hyère saltern	05. 2002	*A. franciscana* (p. s.)	
Thau Castelan saltern	05. 2002	*A. franciscana* (p. s.)	

A. parthenog. (d): *Artemia parthenogenetica* (diploid); (t): tetraploid; (p.s.): alcohol preserved original specimens.

Table 3. Artemia cyst samples available from Italy.

Locality	Sampling date	Cyst diameter	Observations
Veneto province			
Comacchio salterns	1985	278	NH
Apulia province			
Margherita di Savoia salterns	1988	258	*A. parthenogenetica* (d): 67% *A. parthenogenetica* (t): 33%
Margherita di Savoia salterns	1988	267	*A. parthenogenetica* (d): 22% *A. parthenogenetica* (t): 78%
Lacio province			
Tarquinia salterns	2002	243	*A. salina*
Sicilia province			
Isola Longa salterns	1985	245	NH
Sardinia province			
Cagliari salterns	1988	254	*A. salina*
Carloforte salterns	1987	251	NH
Carloforte salterns	1988	256	*A. salina* *A. parthenogenetica* (d) <1%
San Antioco salterns	?	255	*A. salina*
ARC 579			
Santa Gilla salterns	1994	253	*A. salina*

Mean diameter of cysts (μm) measured with a micrometer eyepiece. Relative specific composition where different species or strains appeared. (d): *Artemia parthenogenetica* diploid, (t): tetraploid.
NH: No hatching cysts. ARC 579: *Artemia* Reference Center cyst bank sample.

males, respectively. The populations analysed can be split into three different groups when females (Figure 1) are considered and two different groups for males. Two of the three groups obtained for females are shown to be quite homogeneous, and these include 13 *A. franciscana* and 9 *A. salina* populations. The third group, more complex and less homogeneous, includes diploid and tetraploid parthenogenetic populations.

When males are considered (Figure 2), only two groups are observed owing to the absence of males for parthenogenetic populations. The group concerning *A. salina* males looks more homogeneous, whereas the other group, dealing with *A. franciscana* males, shows a clear split between a group of males morphologically similar to those originally from Great Salt Lake (Utah, USA), whereas the others are similar to San Francisco Bay (California, USA) ones, suggesting the possibility that the different American brine shrimp populations introduced in the Western Mediterranean localities originate from cysts imported from both parts of the USA. These data also support the view of Pilla and Beardmore (1994) on the greater usefulness of male traits in this type of morphological analysis.

Discussion

The first recorded deliberate introductions of *Artemia franciscana* were those carried out on a Pacific Island and in Brazil in the 1970s (VanStappen 2002). According to our results obtained through the screening of old and updated brine shrimp cyst samples collected in salterns from various western Mediterranean countries, including the Atlantic shore salterns in Portugal and southwest Spain, the presence of *A. franciscana* as an exotic invasive species is confirmed in Portugal, Spain, France, as well as in the north of Morocco.

The presence of *A. franciscana* in the southwest of Portugal appears to date from the early 1980s as suggested by previous information (Hontoria et al. 1987). The American brine shrimp populations probably then spread (or was introduced) to the North, invading hypersaline environments in central Portugal, i.e., Sado and Tejo estuaries during the course of this decade, reaching the salterns in the Aveiro district at the end of 1980s or early 1990s and outcompeting autochthonous *Artemia* populations.

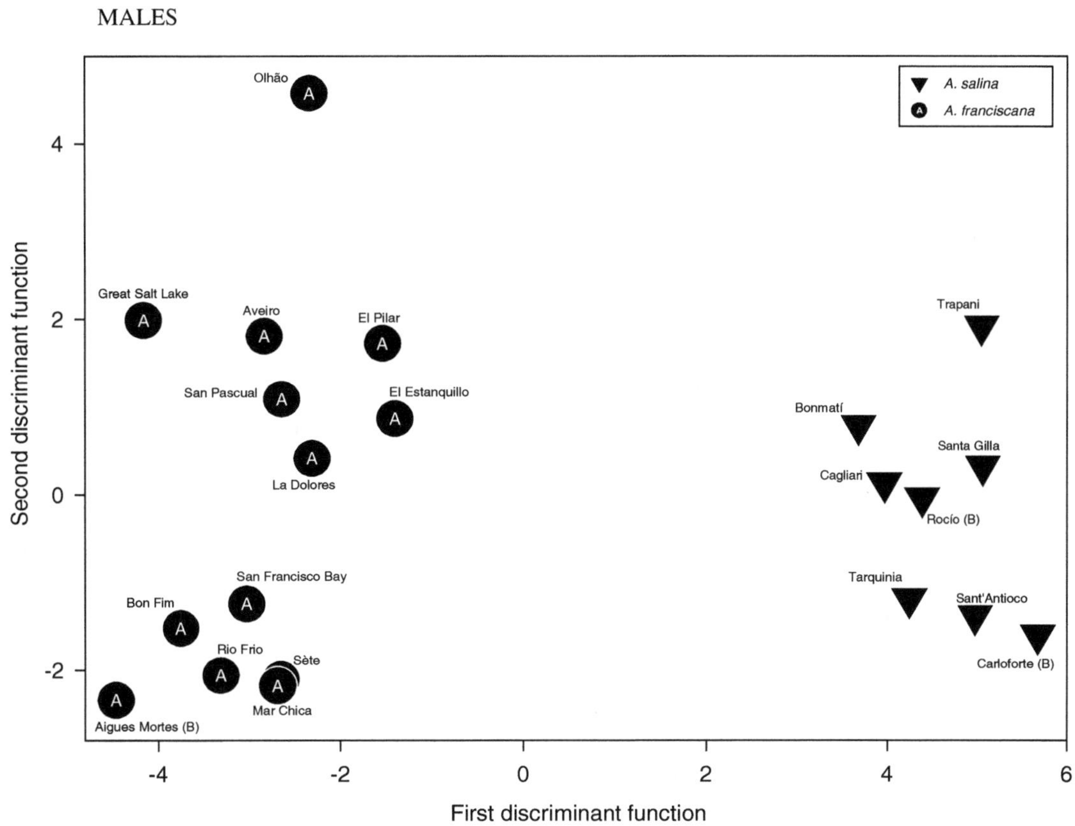

Figure 2. Group centroids of the populations studied for the first two discriminant functions resulting from the discriminant analysis on male morphometric variables.

The salterns in the southwest of Spain, especially in the provinces of Cadiz and Huelva, have been the subject of field studies over several decades. They provide a well-documented case of coexistence of parthenogenetic (diploid and tetraploid) strains and the amphygonic *A. salina* (Amat 1983a; Amat et al. 1995). However, in most of the 140 old salterns exploited around the Cadiz Bay (Puerto Real, San Fernando, Chiclana and Cadiz), salt extracting activity ceased at the beginning of the 1980s (Amat 1983b), their land use changing to aquacultural projects. This fact presumably provoked non-intentional inoculations of *A. franciscana* through hatchery effluents, while a few of them became the source of brine shrimp products for aquarists and pet markets after intentional inoculations (J.A. Calderon, pers. comm.).

The same aquacultural source for the introduction of American brine shrimp in the south of Spain and Portugal is presumably responsible for its presence in the French Mediterranean (see below), where an unusual sexual population of *Artemia* was previously recorded by Thiery and Robert (1992). A close inspection of their paper shows that they had found an introduced population of *A. franciscana*, although they failed to name the species.

Italian samples studied from eight localities do not show the presence of *A. franciscana*. In these Italian salterns, only the autochthonous strains were found. This supports the results of Nascetti et al. (2003) in their study of the genetic structure of Italian *Artemia* from brackish and hypersaline waters.

According to Ehrlich (1984), Lodge (1993) and McMahon (2002), successful invasive species usually display different degrees of the following attributes: (1) abundance in their original range or large native range, (2) polyphagous or eurytrophic (i.e. wide feeding niche), (3) much genetic variability or phenotypic plasticity, (4) short generation times, (5) fertilized females able to colonize alone (i.e. single parent reproduction), (6) vegetative reproduction, (7) larger than most related species, (8) high dispersal rate, (9) associated with human activities (human commensalism) and (10) able to function in a wide range of physical conditions.

In fact, the biological attributes of species are not the only reason of successful introductions. According to Williams (1996) and Blackburn and Duncan (2001), the match between the climatic or environmental conditions in a species natural range and the climate in the location of the introduction could also be an important factor.

There are many examples of invading aquatic species showing most of the cited traits, but not all these traits can play a definitive role in determining the success of invasions, the shifts in the structure of invaded communities over time, and the probability of extinction of autochthonous species. How well *Artemia* complies with most of these traits thought to characterize invasive species is addressed below.

It has been argued that there is little evidence suggesting that physiological capacity to tolerate and function in a wide range of physical conditions is a prerequisite to successful invasions of aquatic habitats (McMahon 2002). In the genus *Artemia*, although the information on the tolerance of its different species to a wide range of physical conditions is not complete, it is commonly accepted that its adaptation to the severe habitats of hypersaline ecosystems, and its wide distribution in all continents except Antarctica, mean that the species and populations can withstand the widest salinity and temperature ranges among aquatic organisms, living in salinities at or below seawater concentration (35 g l^{-1}), up to saturation level (300 g l^{-1}).

It is possible to assimilate the filter feeding mechanism of *Artemia* to a polyphagous regime equivalent to eurytrophy, provided that the filtered particles do not exceed a range of critical sizes, i.e., between 6.8 and 27.5 μm (Gelabert 2001).

The sympatry of sexual and asexual autochthonous *Artemia* species recorded in several sites, e.g. Spanish Mediterranean coastal salterns (Browne and McDonald 1982; Amat 1983a; Browne et al. 1988), or lake Urmia (Iran) and peripheral hypersaline lagoons (N. Agh, pers. comm.) has previously motivated laboratory studies of the interaction between sexual and asexual populations. These competition studies can shed light on life history traits that may explain the success of *A. franciscana* as an invader.

Competition experiments between bisexual and asexual *Artemia* populations from several localities of the Old World have been carried out in the laboratory (Browne 1980; Browne and Halanych 1989), with *A. franciscana* usually incorporated as a model species since it is the best known and studied taxon (Lenz and Browne 1991). In these experiments, *A. franciscana* populations outcompeted parthenogenetic (diploid) populations in 91% of the scorable trials. However, when these *A. parthenogenetica* populations competed against the co-occurring Old World and Mediterranean sexual populations *A. salina*, bisexuals were eliminated in 98% of the trials. Thus, the competitive abilities of the *Artemia* populations under the experimental conditions tested are *A. franciscana* > *A. parthenogenetica* > *A. salina*. However, salinity, temperature and other environmental gradients are likely to influence the relative performance of each species.

Many invasive aquatic species with significant ecological impacts are characterized by adaptations supporting rapid population growth, including rapid individual growth, early maturity, short generation times, high fecundities and small egg–offspring size. These are traits characteristic of species adapted to unstable habitats, with frequent population density reductions or disappearance associated with unpredictable natural environmental events (Browne and Wanigasekera 2001). *Artemia* and especially *A. franciscana* can be considered to exhibit these traits, *A. franciscana* being a more extreme *r*-strategist (Lodge 1993; Williamson 1996) compared to the species it outcompetes.

A wide variety of factors such as environmental cues, life-history traits, heterozygosity levels and genetic variability may contribute in determining the competitive abilities of the *Artemia* populations. These variables cannot be pooled together in experimental designs, but the partial information available from studies of each factor and the evidence from studies in the field where competition occurs suggest the possibility of competitive exclusion of native *Artemia* by

A. franciscana in 10–100 generations (Miller 1967), although according to Lenz and Browne (1991) it may be attainable in two or three generations.

Most *Artemia* species reproduce primarily by ovoviviparity under favourable environmental factors, especially when there is no food limitation, but they switch to oviparity (producing cysts that can undergo prolonged diapause and cryptobiotic periods) when unfavourable environmental factors threaten the ovoviviparous reproduction and the population persistence. These cysts are the best way to ensure the appearance of a new population under renewed favourable conditions promoting cyst hydration and hatching, i.e. the renewal of the population in the following season. These cysts are also the best way to ensure a successful dispersion of *Artemia* populations. These cysts are thought to be broadly dispersed by wind transportation among short distances, or by birds for longer distances (Figuerola and Green 2002; Green et al. 2002).

Last but not the least, the presence of *A. franciscana* from the New World in the Western Europe and Mediterranean shores is unquestionably associated with a human activity, with the aquaculture of marine species of commercial interest. With the unavoidable development of hatcheries to obtain postlarvae and fingerlings for aquaculture, the use of *Artemia* nauplii as a diet for larval culture became widespread. Dormant cysts of *Artemia* can be stored for long periods in cans and then used as an off-the-shelf food requiring only 24 h of incubation to obtain live nauplii (Lavens and Sorgeloos 2000). From the early 1950s, commercial sources of cysts initially originated from the coastal salterns in the San Francisco Bay, California, USA, and the inland Great Salt Lake, Utah, USA. These cysts were primarily marketed for the aquarium pet trade, and in the mid-1970s, the demand increased from emerging aquaculture operations, currently attaining requirements of about 2000 metric tons of cysts annually.

Shrimp culture, based on Penaeids, started to develop rapidly in the Mediterranean area (mainly in Italy and France) in the latter years of the 1970 decade (Lumare 1990). Experimental productions of *Penaeus japonicus* in the southwest of Spain (Cadiz) started in 1982, and in 1986, about 20 million postlarvae were reared (Rodriguez 1986). Shrimp culture was largely replaced by marine finfish (sea bass, sea bream) culture in the early 1980s, when many old seaside salterns that had been traditionally exploited in the area of Cadiz bay switched their activity to intensive fish culture. By 1998, 900 ha of old saltern ponds had been converted to intensive fish culture (Espinosa et al. 1999).

In southwestern Portugal (Algarve province), old salterns were converted to prawn and fish culture in 1985–1986 (Gouveia 1994), but American brine shrimp may have been introduced earlier for the aquarium pet trade (M.N. Vieira, pers. comm.).

In the early 1970s, there were important advances in sea bass intensive cultures in pilot plants settled in the area of Sete (Languedoc, France) near coastal lagoons and marine salt exploitations (Barnabe 1974a, b). During this decade, several fish farms developed in the lagoons and brackish environments along the Languedoc shore: Salses-Leucate, Thau., etc.

These aquaculture developments have been closely linked to the success in shrimp and fish hatchery productions, where larviculture relied on the supply of live food organisms in sufficient quantity. To date, these living preys are rotifers (*Brachionus plicatilis*) obtained through season-round massive culture, and *Artemia* nauplii from the massive hatching of cysts purchased in the international market from sources in America, especially from Great Salt Lake, Utah.

Our results suggest that the native populations of *Artemia* in the Mediterranean region are under severe threat from competition with the expanding *A. franciscana*. In an attempt to prevent or slow down further spread of this exotic species, we suggest that aquaculture activities should be subject to tighter regulation. Where possible, the use of cysts from native species should be encouraged.

Acknowledgements

This study has been funded by the Spanish Government R & D National Plan under the projects AGL 2001-1968, INCO Project ICA4-CT-2002-10020 and AGL 2001-4582 E. Olga Ruiz has been supported by a fellowship of the FPU Programme from the Spanish Ministry of Education and Culture. We would like to thank Claudine de la Court and Nuno Grade for their help in collecting cyst samples. The manuscript incorporated the suggestions of an anonymous reviewer which helped improve it.

References

Abatzopoulos ThJ, Zhang B and Sorgeloos P (1998) International study on *Artemia*. LIX. *Artemia tibetiana* preliminary characterization of a new *Artemia* species found in Tibet (People's republic of China). International Journal of Salt Lake Research 7: 41–44

Abatzopoulos ThJ, Kappas I, Bossier, P, Sorgeloos P and Beardmore JA (2002) Gene characterization of *Artemia tibetiana* (Crustacea, Anostraca). Biological Journal of the Linnean Societry 75(3): 333–344

Abreu-Grobois FA and Beardmore JA (1982) Genetic differentiation and speciation in the brine shrimp *Artemia*. In: Barigozzi C (ed) Mechanisms of Speciation, pp 345–376. Alan Liss, Inc., New York

Amat F (1983a) Diferenciación y distribución de las poblaciones de *Artemia* de España, VI. Biogeografía. Investigación Pesquera 47(2): 231–240

Amat F (1983b) Zygogenetic and parthenogenetic *Artemia* in Cadiz sea-side salterns. Marine Ecology – Progress Series 13: 291–293

Amat F, Barata C, Hontoria F, Navarro JC, and Varó I (1995) Biogeography of the genus *Artemia* (Crustacea, Branchiopoda, Anostraca) in Spain. International Journal of Salt Lake Research 3(2): 175–190

Artom C (1906) La variazione dell*Artemia* salina (Linn.) di Cagliari sotto l'influsso della salsedine. Atti della R. Accademia delle Scienze di Torino. XLI: 971–972

Badaracco G, Baratelli L, Ginelli E, Meneveri R, Plevani P, Valsasnini P and Barigozzi C (1987) Variations in repetitive DNA and heterochromatin in the genus *Artemia*. Chromosoma 95: 71–75

Barnabe G (1974a) Compte rendu sommaire de la campagne 1972–1973 de reproduction contrôlée du loup à Sète. Publ. CNEXO (Actes et Colloques) 4: 205–213

Barnabe G (1974b) Mass rearing of the bass Dicentrarchus labrax L. In: Blaxter JHS (ed) The Early Life History of Fish, pp 749–753. Springer-Verlag, Berlin

Blackburn TM and Duncan RP (2001) Determinants of establishment success in introduced birds. Nature 414: 195

Browne RA (1980) Reproductive pattern and mode in the brine shrimp. Ecology 61: 466

Browne RA and Halanych KM (1989) Competition between sexual and parthenogenetic *Artemia*: a re-evaluation (Branchiopoda, Anostraca). Crustaceana 57: 57–71.

Browne RA and MacDonald GH (1982) Biogeography of the brine shrimp *Artemia*: distribution of parthenogenetic and sexual populations. Journal of Biogeography 9: 331–338

Browne RA and Wanigasekera G (2000) Combined effects of salinity and temperature on survival and reproduction of five species of *Artemia*. Journal of Experimental Marine Biology and Ecology 244: 29–44

Browne RA, Davis LE and Sallee SE (1988) Effects of temperature and relative fitness of sexual and asexual brine shrimp *Artemia*. Journal of Experimental Marine Biology and Ecology 124: 1–20

Browne RA, Li M, Wanigasekara G, Simonek S, Brownlee D, Eiband G and Cowarn J (1991) Ecological, physiological and genetic divergence of sexual and asexual (diploid and polyploid) brine shrimp *Artemia*. Advances in Ecology 1: 41–52

Cai Y (1989) A redescription of the brine shrimp (*Artemia sinica*). The Wasmann Journal of Biology 47: 105–110

Ehrlich PR (1984). Which animal will invade? In: Mooney HA and Drake JA (eds) Ecology of Biological Invasions of North America and Hawaii, pp 79–95. Springer-Verlag, New York

Espinosa J, Diaz V, Labarta U, Muñoz E, Toribio MA and Ruiz A (1999) La investigación y el desarrollo tecnológico de la acuicultura en España en el período 1982–1997. Ministerio de Agricultura, Pesca y Alimentación, 136 pp

Figuerola J and Green AJ (2002) Dispersal of aquatic organisms by waterbirds: a review of past research and priuorities for future studies. Freshwater Biology 47: 483–494

Gajardo G, Abatzopoulos ThJ, Kappas I and Beardmore JA (2002) Evolution and Speciation. In: Abatzopoulos ThJ, Beardmore JA, Clegg JS and Sorgeloos P (eds) *Artemia*. Basic and Applied Biology, pp 225–250. Kluwer Academic Publishers, Dordrecht, The Netherlands

Gelabert R (2001) Artemia bioencapsulation I. Effect of particle sizes on the filtering behaviour of *Artemia franciscana*. Journal of Crustacean Biology 21(2): 435–442

Gilchrist BM (1960). Growth and form of the brine shrimp *Artemia salina* (L). Proceedings of the zoological society of London 431(2): 221–235

Gouviea A (1994) Aquaculture in Portugal, state of the art, constraints and perspectives. Aquaculture 19(1): 23–28

Green AJ, Figuerola J and Sánchez MI(2002) Implications of waterbird ecology for the dispersal of aquatic organisms. Acta Oecologica 23: 177–189

Günther RT (1890) Crustacea. In: Günther RT (ed) Contributions to the Natural History of Lake Urmi, N.W. Persia and its Neighbourhood. Journal of the Linnean Society (Zoology) 27: 394–398

Hontoria F (1990) Caracterización de tres poblaciones originarias del área levantina española del crustáceo branquiópodo *Artemia*. Aplicación en Acuicultura Tesis, Universidad Autónoma de Barcelona, 326 pp

Hontoria F and Amat F (1992a) Morphological characterization of adult *Artemia* (Crustacea, Branchiopoda) from different geographical origin. Mediterranean populations. Journal of Plankton Research 14(7): 949–959

Hontoria F and Amat F (1992b) Morphological characterization of adult *Artemia* (Crustacea, Branchiopoda) from different geographical origin. American populations. Journal of Plankton Research 14(10): 1461–1471

Hontoria F, Navarro JC, Varó I, Gozalbo A, Amat F and Vieira MN (1987) Ensayo de caracterización de cepas autóctonas de *Artemia* de Portugal. Seminario Aquacultura Instituto Ciências Biomédicas 'Abel Salazar', Porto, Portugal, 10 pp

Hsü K, Montadert L, Bernoulli D, Cita MB, Erickson A, Garrison RE, Kidd RB, Melieres F and Müller C (1977)

History of the Mediterranean salinity crisis. Nature 267: 399–403
Kellogg VA (1906) A new *Artemia* and its life conditions. Science 24: 594–596
Lavens P and Sorgeloos P (1987) The cryptobiotic state of *Artemia* cysts, its diapause deactivation and hatching: a review. In: Sorgeloss P, Bengtson DA, Decleir W and Jaspers E (eds) *Artemia* Research and Its Applications, Vol 3, pp 27–63. Universa Press, Wettern, Belgium
Lavens P and Sorgeloos P (2000) The history, present status and prospects of the availability of *Artemia* cysts for aquaculture. Aquaculture 181: 397–403
Lenz PH and Browne RA (1991) Ecology of *Artemia*. In: Browne RA, Sorgeloos P and Trotman CNA (eds) *Artemia* Biology, pp 237–253. CRC Press, Boca Raton, Florida
Lodge DM (1993) Biological invasions: lessons for ecology. Trends in Ecology and Evolution 8: 133–137
Lumare F (1990) Crucial points in research into and commercial production of shrimps. In: Flos R, Tort L, and Torres P (eds) Mediterranean Aquaculture, pp 21–40. Ellis Horwood, Aberdeen, UK
McMahon RF (2002) Evolutionary and physiological adaptations of aquatic invasive animals: *r* selection versus resistance. Canadian Journal of Fish Aquatic Sciences 59: 1235–1244
Miller RS (1967) Patterns and process in competition. Advances in Ecology Research 4: 1–10
Narciso L (1989) The brine shrimp *Artemia* sp.: an example of the danger of introduced species in aquaculture. EAS Special Publication 10: 183–184
Nascetti G, Bondanelli P, Aldinucci A and Cimmaruta R (2003) Genetic structure of bisexual and parthenogenetic populations of *Artemia* from Italian brackish-hypersaline waters. Oceanologica Acta 26: 93–100
Piccinelli M and Prosdocimi T (1968) Descrizione tassonomica delle due species *Artemia salina* L. e *Artemia persimilis* n. sp. Istituto Lombardo, Accademia di Scienze e Letter, Rendiconti B 102: 170–179
Pilla EJS and Beardmore JA (1994) Genetic and morphometric differentiation in Old World *Artemia*, PhD Thesis, University of Wales, Swansea, UK
Rodriguez A (1986) Prawn culture in Spain: status and problems. International Workshop on Marine Crustacean Culture, Development and Management, Venice, pp 1–6
Stefani R (1960) L'*Artemia* salina partenogenetica a Cagliari. Rivista di Biologia, Perugia LII: 463–491
Stella E (1933) Phenotypical characteristics and geographical distribution of several biotypes of *Artemia salina*. Zeitschrift fur Inductive Abstammungs and Vererbungslehre 65: 412–446
Thiery A and Robert F (1992) Bisexual populations of the brine shrimp *Artemia* in Sète-Villeroy and Villeneuve saltworks (Languedoc, France). International Journal of Salt Lake Research 1: 47–63
Tryantaphyllidis GV, Criel GRJ, Abatzopoulos TJ and Sorgeloos P (1997a) International Study on *Artemia*. LIV. Morphological study of *Artemia* with emphasis to Old World strains. II. Parthenogenetic populations. Hydrobiologia 357: 155–163
Tryantaphyllidis GV, Criel GRJ, Abatzopoulos TJ, Thomas KM, Peleman J, Beardmore JA and Sorgeloos P (1997b) International Study on *Artemia*. LVII. Morphological and molecular characters suggest conspecificity of all bisexual European and North African *Artemia* populations. Marine Biology 129: 477–487
Vanhaecke P and Sorgeloos P (1980) International Study on *Artemia*. IV. The biometrics of *Artemia* strains from different geographical origin. In: Persoone G, van Sorgeloos P, Roels O and Jaspers E (eds) The Brine Shrimp *Artemia*, pp 393–405. Universa Press, Wetteren, Belgium
Vanhaecke P, Tackaert W and Sorgeloos P (1987) The biogeography of *Artemia*: an updated review. In: Sorgeloos P, Bengtson DA, Decleir W, and Jaspers E (eds) *Artemia* Research and its Applications, Vol 1, pp 129–155. Universa Press, Wetteren, Belgium
VanStappen G (2002) Zoogeography. In: Abatzopoulos TJ, Beardmore JA, Clegg JS, and Sorgeloos P (eds) *Artemia*: Basic and Applied Biology, pp 171–224. Kluwer Academic Publishers, Dordrecht, The Netherlands
VanStappen G, Lavens P and Sorgeloos P (1998) Effects of hydrogen peroxide treatment in *Artemia* cysts of different geographical origin. Archiv für Hydrobiologie. Special Issues Advanced Limnology 52: 281–296
Vieira MN (1990) Contribuiçâo para o conhecimento da biología de *Artemia* sp. proveniente das salinas de Aveiro. Sua importância en Aquacultura e na dinâmica daquele ecossistema. Tese de Doutoramento. Faculdade de Ciencias do Porto, Portugal 324 pp
Vieira MN and Amat F (1985) *Artemia* sp. from Aveiro: its characterization. Publicaçoes do Instituto de Zoologia 'Dr. Augusto Nobre'. Faculdade de Ciencias do Porto 191: 1–9
Williamson MH (1996). Biological Invasions. Chapman & Hall, New York

Biological Invasions (2005) 7: 49–73

Impact of an introduced Crustacean on the trophic webs of Mediterranean wetlands

Walter Geiger*, Paloma Alcorlo, Angel Baltanás & Carlos Montes
*Departamento de Ecología, Universidad Autónoma de Madrid, 28049 Madrid, Spain; *Author for correspondence (e-mail: walter.geiger@uam.es; fax: +34-91-3978001)*

Received 4 June 2003; accepted in revised form 30 March 2004

Key words: ecological impact, invasive species, Mediterranean wetlands, *Procambarus clarkii*, trophic web

Abstract

Based on a review and our own data, we present an overview of the ecological impacts on the trophic web of Mediterranean wetlands by an introduced Decapod Crustacean, the red swamp crayfish (*Procambarus clarkii*). *P. clarkii* lacks efficient dispersal mechanisms but is very well adapted to the ecological conditions of Mediterranean wetlands (fluctuating hydroperiods with regular intervals of drought). As an opportunistic, omnivorous species, which adapts its ecology and life history characteristics, such as timing and size at reproduction to changing environmental conditions, it became readily established in most of the Mediterranean wetland environments. High reproductive output, short development time and a flexible feeding strategy are responsible for its success as an invader. Like most crayfish, it occupies a keystone position in the trophic web of the invaded system and interacts strongly with various trophic levels. It efficiently grazes on macrophytes and is one of the main factors, besides the impact of flamingos, cattle and introduced fish, of the change of many water bodies from a macrophyte dominated, clear water equilibrium to a phytoplankton driven turbid water balance. Juveniles feed on protein rich animal food with the corresponding impact on the macroinvertebrate community in competition with other crayfish or fish species. At the same time, it serves as a prey for mammals, birds and fish. Due to its predatory and grazing activity, it efficiently canalises energy pathways reducing food web complexity and structure. Feeding also on detritus it opens, especially in marshlands, the detritic food chain to higher trophic levels which results in an increase of crayfish predators. As a vector of diseases, it has a severe impact on the preservation and reintroduction of native crayfish. *P. clarkii* accumulates heavy metals and other pollutants in its organs and body tissues and transmits them to higher trophic levels. Due to the long history of its presence, the complex interactions it established within the invaded ecosystems and the socio-economic benefits it provides to humans, prevention and control seem the most promising management measures to reduce the negative impact of this crayfish species.

Introduction

Biological invasions and their negative impact on resident communities and ecosystem functioning are considered one of the major threats to biodiversity. Mediterranean ecosystems in particular have a long history of biological invasions be they anthropogenic or non-anthropogenic in origin (Di Castri 1990). Especially threatened are Mediterranean wetlands, which have to suffer the consequences of invasions and of anthropogenic alterations and transformations leading to habitat destruction and very often to their complete disappearance. Despite that these wetlands are second only to rainforests as reservoirs of biodiversity and productivity and are ranked second to estuaries in terms of ecosystem services provided to human welfare (Costanza et al. 1997) they have become only recently the object of increased protection.

Wetland ecosystems are characterised by high biodiversity and complex trophic interactions. Such systems are thought to be less vulnerable to invasions (Sakai et al. 2001), but recent studies have shown that the length of disturbance-free periods is equally important (Shea and Chesson 2002). Disturbance tends to disrupt existing interaction among species and opens new niches for potential invaders. Levels of both anthropogenic and non-anthropogenic disturbances in wetlands are high. A further characteristic of Mediterranean wetlands is the existence of frequent, regular periods of drought, which protects them against most of the invaders.

The impact of an invader also depends on its position in the trophic web of the invaded ecosystem. Species with strong interactions or which are keystone species in the sense of Power and Tilman (1996) will have a larger impact than species with weak or few interactions. Equally, species interacting with several trophic levels affect ecosystem structure and function more intensively than those which interact with a single trophic level. The removal of a species which has already established tight trophic links with native species might produce unpredictable secondary effects on the invaded community. Therefore, an understanding of the invader's role within the trophic web is crucial not only for predictive purposes but also for estimating the consequences of management measures.

Crayfish have been introduced in many water bodies for a long time. Omnivorous and highly active, they are known to occupy keystone positions in both their natural and host ecosystems (Holdich 2002). Therefore, the impact and changes they cause on natural ecosystems once introduced are expected to be high. Nevertheless, they lack efficient systems of dispersal such as easily transported resting eggs or highly mobile larval stages, and their natural potential of dispersal is low in comparison to plants or invertebrate species such as insects or molluscs. However, man has played a crucial role in helping crayfish to overcome this disadvantage by continuous translocations across natural boundaries. Once translated, crayfish establish stable populations followed by rapid range expansion within the invaded watershed.

In what follows, we will try to give an overview based on a literature review and our own studies on the manyfold impacts of an introduced crayfish species – the red swamp crayfish *Procambarus clarkii* – on the natural ecosystems of Mediterranean wetlands.

The biological basis of invasiveness – the example of the red swamp crayfish *Procambarus clarkii*

Successful invaders are characterised by a number of biological and ecological features determining both the process of dispersion and the establishment in the new habitat (Table 1). Although most likely none of the species possesses all of these traits, it is evident that the more they have these traits, the higher their invasive potential is. In the case of *P. clarkii*, not all of these characteristics are equally well expressed. Natural dispersal ability across drainage basins is low, despite the mobility of adults, but this handicap is largely offset by human transport. Although it does not reproduce asexually nor parthenogenetically – but see the recent description of a close parthenogenetic relative in Germany (Scholtz et al. 2002) –

Table 1. Biological and ecological characteristics of successful invaders (Baker 1974) shared by *P. clarkii* (– absent; + low; ++ medium; +++ high).

Biological characteristics of invaders	*Procambarus clarkii*
High dispersal capability through seeds, eggs or highly mobile larval stages	+
Ability to reproduce both sexually and asexually	–
High fecundity	++
Short generation and juvenile development times	++
Fast adaptation to environmental stress	+++
High tolerance to environmental heterogeneity	+++
Desirability to and association with humans (edibility, game species)	+++
Additional features	
Omnivory	+++
Brood care	+++

high reproductive investment of both males (spermatophore production) and females (high egg numbers) increases reproductive success (Gherardi 2002). The species is amongst the most prolific crayfish with more than 600 eggs/females. It reproduces more than once per year if conditions are favourable and adapts its size at maturity to environmental conditions (hydroperiod, food conditions). Newly hatched juveniles are carried by their mothers during the period where they are most vulnerable to predation and reach maturity within several months.

Environmental conditions in the home area of *P. clarkii* are similar to those encountered in Mediterranean wetlands both characterised by regular periods of drought, and this species is very well adapted to withstand these periods in burrows, where they also bear their offspring.

P. clarkii is an opportunistic, omnivorous feeder which readily accepts new food items another advantage when arriving in a new habitat. For these characteristics, which result in easy culturing and high yields, it is prized by humans as a food source, used for baiting and as a laboratory animal and pet. Therefore, it is not astonishing that such a productive species is also a successful invader.

The history of introduction and expansion of *P. clarkii* in Europe and Spain

Crayfishing in Europe for human consumption has been a deep-rooted habit in most parts of the continent. For this reason, traditional management of native crayfish populations through additions and translocations of native species was common. Because overexploitation of this resource extinguished some of the populations, the introduction of exotic species during the XIX century was considered as a possible solution to restore crayfish populations (Lodge et al. 2000a). At least seven species of non-native crayfish have been introduced in Europe since then: five of them were introduced from North America (*Pascifastacus leniusculus, Orconectes limosus, O. immunis, Procambarus clarkii, P. zonangulus*), one from Australia (*Cherax destructor*) and, finally, one from eastern Europe (*Astacus leptodactylus*) (Hobbs 1988; Diéguez-Uribeondo 1998).

But overexploitation by fishing for recreational or commercial purposes is not the only cause of the dramatic decrease of native European crayfish populations, which led in some cases to their extinction. Anthropogenic alteration of river ecosystem quality due to contamination, the alteration of riverine vegetation or riverbed dredging (Alderman and Polglase 1988; Taugbol et al. 1993) the introduction of exotic species, carriers of diseases (Smith and Söderhäll 1986; Taugbol and Skurdal 1993; Diéguez-Uribeondo et al. 1997; Holdich 1997, 1999a) and competitors of native crayfish for shelter and food (Hill and Lodge 1999) contributed substantially to the decline of native crayfish.

One of the most widespread diseases carried by introduced crayfish from North America is a fungal plague called *aphanomicosis*, produced by the oomycete fungus *Aphanomyces astaci*, which is endemic to many North American crayfish but lethal to European crayfish (Unestam 1972; Diéguez-Uribeondo et al. 1995; Alderman 1996). Ironically, the extirpation of native European crayfish by the plague has increased the number of subsequent introductions of North American crayfish (*Orconectes limosus, O. immunis, Pacifastacus leniusculus* and *Procambarus clarkii*) into more than 20 European countries to replace the native stocks (Lodge et al. 2000a).

The introduction of red swamp crayfish (*Procambarus clarkii*), the subject of this review, in Europe is a very well documented example of the quick expansion of an alien species. It was first introduced in 1973 in Spain in two aquaculture installations located in Sevilla (Lower Guadalquivir River Basin, southwestern Spain) and Badajoz (southwestern Spain) (Habsburgo-Lorena 1983). The aim of the introduction was twofold: On the one hand, there were economic arguments; it was an attempt to improve the economy of an impoverished area by developing crayfish commercialisation plans. On the other hand, it was erroneously thought that the introduction of a non-native species into an area without native crayfish would cause no ecological problems, because the red swamp crayfish would occupy a new empty niche. The fact is, that, in only three decades, red swamp crayfish became widespread throughout the Mediterranean region and Europe. Several factors, all of them linked to human activity such as the increasing economic importance of *P. clarkii*, its *in vivo* commercialisation and repeated translocations for economical or recreational purposes, are responsible for its rapid spread. From southwestern Spain, *P. clarkii* populations expanded to the rest of the country including the Balearic (Majorca: (Hobbs et al. 1989)) and Canary Islands (Gutiérrez-Yurrita and Martínez 2002) and to Europe: Portugal (Ramos and Pereira 1981; Correia 1992; Adao and Marques 1993), Azores Islands (Correia and Costa 1994), Cyprus (Hobbs et al. 1989), United Kingdom (Holdich 1999b), France (Arrignon et al. 1999), Italy (Gherardi et al. 1999), Netherlands (Hobbs et al. 1989) and Switzerland (Stucki 1997; Stucki and Staub 1999).

Effects of alien crayfish in food webs – general aspects

In many ecosystems, crayfish occupy a central position in the trophic web acting as both predator and prey. As opportunistic, omnivorous feeders, they include in their diet submerged macrophytes, algae, invertebrates and detritus (Lodge and Hill 1994; Momot 1995; Gutiérrez-Yurrita et al. 1998). In the words of Huner (1981), 'They eat any insect, crustacean, molluscs (especially snails), or annelid worm they can catch.'

Invasive crayfish species clear macrophyte beds thereby altering the ecosystem characteristics such as habitat heterogeneity (Lodge and Lorman 1987; Lodge et al. 2000b) or the composition of invertebrates associated with macrophytes. In addition, they feed directly on many invertebrate species, reducing their abundances (Nyström et al. 1996; Perry et al. 1997).

Crayfish diet is reported to change with body size. Small crayfish are mainly carnivorous, and larger individuals are primarily herbivorous (Abrahamsson 1966; Lorman and Magnuson 1978). This ontogenetic shift has also been observed in red swamp crayfish. Animal food is much more important for young, rapidly growing juveniles than for adults (Marçal-Correia 2003). Since crayfish cannot swim, foraging they concentrate on the bottom or benthic zone. However, some individuals, especially young ones, can catch planktonic organisms with their mouth parts acting as a filter. Living green plant material, an important source of dietary carotenoids (Huner 1981), also forms part of the red crayfish's diet. Other studies postulate that the principal food of the red crayfish is plant detritus (Lorman and Magnuson 1978). Once dead, submerged plants quickly become covered with a layer of living bacteria and fungi which use the dead plant material as an energy source. The dead plant material itself is of little energetic value to the red crayfish, but not so the rich protein layer of bacteria and fungi (Cronin 1998).

Besides these effects on lower trophic levels (top-down effect), they also serve as a prey to higher trophic levels (bottom up effect), and they are also known to compete with fish and other crayfish species for food (Momot 1995).

In the following chapters, we would like to examine in detail the impact of the introduced red swamp crayfish on the different trophic levels of the invaded ecosystems

Impacts of *P. clarkii* on macrophytes

Crayfish feeding and macrophytes

Several studies have demonstrated that crayfish consume freshwater macrophytes, with plants often accounting for over 75% of the diet (King 1883; Chidester 1908; Tack 1941; Momot 1967; Prins 1968). They are common and important omnivores which consume a lot of living plant tissue and detritus when favoured animal prey is not available (Momot 1995). Crayfish can reduce (Abrahamsson 1966; Rickett 1974; Saiki and Tash 1979; Carpenter and Lodge 1986; Feminella and Resh 1986), or eliminate submerged vegetation from the littoral zone of many lakes and ponds whether they are native (Dean 1969) or have been introduced (Lorman and Magnuson 1978, Chambers et al. 1990). Some species of crayfish are also considered to be large-bodied grazers with both low numerical and biomass density and large effects on filamentous alga (*Cladophora*). Grazer exclusion experiments with large *Orconectes propinquus* resulted in an algae biomass increase of an order of magnitude (Creed 1994). Little quantitative information exists about the relationship between introduced crayfish species density and macrophyte biomass or species composition (Appendix 1). However, some conclusions can be derived from biomanipulation experiments conducted in mesocosms – for example, that crayfish consumption of submerged macrophytes is species-selective and also density-dependent (Lodge and Lorman 1987; Chambers and Hanson 1990). Several authors report that the impact on macrophytes depend on crayfish density (Flint and Goldman 1975; Lodge and Lorman 1987). Chambers et al. (1990) manipulated sex ratios and densities of *Orconectes virilis* to show that macrophyte species are differentially affected by crayfish attack. Furthermore, their observations indicated that macrophyte attack is indiscriminate but that crayfish feeding is selective (Chambers et al. 1990).

In general, the impact of crayfish feeding on macrophytes depends on a combination of three factors: the type of macrophyte (e.g. differences between species, initial biomass, growth form, palatability), the crayfish (e.g. differences between species, sexes, individual crayfish size and activity), and the abundance of alternative prey.

The role of P. clarkii

The dominant herbivorous feeding character of *Procambarus clarkii* has been documented in life history studies from their natural habitats in Louisiana (USA) (Penn 1943; Avault et al. 1983). But so far, little is known about the quantitative effects of this species on macrophytes once introduced elsewhere. Exceptions are the results of crayfish exclusion experiments and submerged macrophytes, performed *in situ*, in the freshwater marshes of Coyote Hills (California, USA), by Feminella and Resh (1986). They found that the exclusion of crayfish resulted in a sixfold increase in macrophytes and that crayfish abundance is strongly related to *Potamogeton pectinatus* clearance (Feminella and Resh 1989).

In multispecies laboratory experiments, Cronin (1998) found that red swamp crayfish avoided macrophyte species with structural or chemical deterrents and preferred undefended plants high in nitrogen. Plant structure (morphology, toughness, and/or surface features) and plant chemistry were important determinants of crayfish feeding choices (Cronin 1998; Cronin et al. 2002).

In Mediterranean environments, *P. clarkii* has been cited to be responsible for the disappearance of some macrophyte species in wetlands and fresh and brackish water marshes of southern Europe (Montes et al. 1993). In Spain, there are other examples where the composition of submerged macrophytes changed following the arrival of *P. clarkii* (e.g. Laguna de El Portil in Huelva, SW Spain (Enríquez et al. 1987); Lake Carucedo (Dpt. Ecología, UAM, unpubl.), Lake Chozas, León, northwestern Spain (Palacios and Rodríguez 2002)). However, at least in freshwater marshes, other factors such as the anthropogenic alteration of water quality and flooding regime or livestock trampling and flamingo treading (Duarte et al. 1990; Montes and Bernués 1991; Grillas et al. 1993) seem to have contributed to the decrease of macrophyte populations in this area. Livestock and flamingos directly damage the macrophyte seed bank (Montes and Bernués 1991), whereas crayfish has a lower impact on this important reservoir of macrophyte diversity.

Shredding of plants and bioturbation by *P. clarkii* are thought to be responsible for the change from a natural, macrophyte dominated, transparent water state equilibrium to a turbid, eutrophic balance, dominated by phytoplankton (Duarte et al. 1990; Nyström and Strand 1996). Angeler et al. (2001) showed in the Tablas de Daimiel wetland of La Mancha, Central Spain, that the benthic feeding of crayfish disturbs and resuspends the sediment, which leads to increased nutrient release. This results in a deterioration of water quality, increased turbidity and nutrient content and reduced light availability for submerged macrophytes. However, importance for nutrient recycling at the ecosystem level was found to be low (Angeler et al. 2001).

Rodríguez et al. (2002) described the disappearance of seven species of submerged macrophytes of a small lake in northwestern Spain (Lake Cabañas, León) after the introduction of *P. clarkii* in 1997. The recovery level of macrophytes in an exclusion experiment was 70%.

Additional quantitative studies directed towards the question on how crayfish density and population structure affect macrophytes and towards the role of other factors (nutrient enrichment, hydroregime changes) should clarify the role of this crayfish species in the disappearance and alteration of macrophyte communities.

The impact of alien crayfish on the native invertebrate communities

P. clarkii as an invertebrate predator

For a long time, crayfish were described to be mainly herbivores and detritivores as gut content analyses always contained large amounts of plant material and detritus (Webster and Patten 1979; Huryn and Wallace 1987). However, when correcting gut contents for assimilation efficiencies, the importance of animals as an energy source increases (Whiteledge and Rabeni 1997). Animals form, at least in the juvenile stage, when growth rates are the highest, an important part of a crayfish's diet (Hobbs 1993; Gutiérrez-Yurrita et al. 1998). A direct impact on its prey organisms is therefore to be expected. Crayfish feed mainly on aquatic invertebrates, with a clear preference for arthropods and gastropods (see for a review Momot 1995). The reduction of invertebrate populations by crayfish feeding has often cascading effects on lower trophic levels. In preference experiments, *P. clarkii* prefers animal food over macrophytes (Ilhéu and Bernardo 1993), whereas in the field, it mainly feeds on plant material and detritus (Feminella and Resh 1986, 1989; Gutiérrez-Yurrita et al. 1998).

Gutiérrez-Yurrita et al. (1998) showed that despite high occurrences of plant material and detritus, small arthropods (copepods, ostracods), insect larve and fish (*Gambusia holbrooki*) are consistently found in the guts. Fish is eaten only by large, adult individuals, whereas copepods are an important food source for small crayfish. Furthermore, these authors observed cannibalism in 20% of the larger sized (>30 mm carapax length) individuals. No differences were found in the feeding preferences of males and females.

Comparing rice fields and natural marshland ecosystems, we were able to demonstrate that crayfish feeding is highly flexible and is a function of prey availability in the field. Crayfish from natural marshlands fed on 17 different prey items, whereas the guts of individuals from rice fields only contained 12 taxa (Figure 1). In both systems, they mainly feed on macrophytes (>97% of occurrence), but the percentage of stomachs with animal food can be as high as 50% in natural marshlands, especially in spring, when prey diversity is the highest. Rice fields are characterised by an impoverished invertebrate fauna, and crayfish fulfill their need for animal protein by cannibalism and predation on fish (Table 2) (Alcorlo et al. in press). Predation on mosquito fish (*Gambusia holbrooki*), which occurs in high

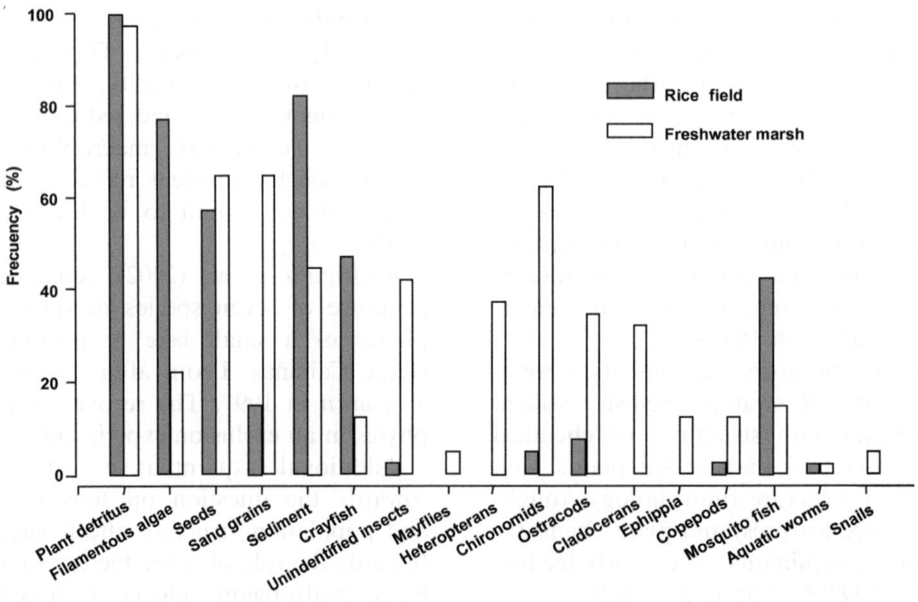

Figure 1. Frequency of occurrence of food items found in guts of *P. clarkii* from rice fields and marshlands of the Lower Guadalquivir basin (Spain).

Table 2. Frequency of occurrence of food items in gut contents of *P. clarkii*.

Food items	Frequency of appearance	
	Rice field	Freshwater marsh
Plant debris	100	97.5
Filamentous algae	77.5	22.5
Seeds	57.5	65
Sand grains	15	65
Clay particles	82.5	45
Crayfish	47.5	12.5
Nonidentified Insects	2.5	42.5
Ephemeroptera	0	5
Heteroptera	0	37.5
Chironomidae	5	62.5
Ostracoda	7.5	35
Cladocera	0	32.5
Ephippia	0	12.5
Copepoda	2.5	12.5
Gambusia holbrooki	42.5	15
Oligochaeta	2.5	2.5
Gastropoda	0	5

densities in the rice fields, is noteworthy as fish is not commonly found among crayfish prey.

P. clarkii *and the extinction of macroinvertebrates including native crayfish*

As stated above, *P. clarkii* is thought to be responsible for the disappearance of some species of macroinvertebrates in aquatic ecosystems where it was introduced. An example is the coincidence between the extinction of two species of gastropods – *Lymnaea peregra* and *L. stagnalis* – in freshwater marshes of the Doñana National Park (southwestern Spain) and the introduction of the red swamp crayfish (Montes et al. 1993). Gastropods are known to be one of the favourite food items in the diet of crayfish (Covich 1977; Goddard 1988; Hanson et al. 1990; Olsen et al. 1991; Ilhéu and Bernardo 1993). *P. clarkii* has been introduced in Kenya to reduce snail populations and thereby snail-born diseases (Rosenthal et al. 2001). It has also been proposed as a control agent of the giant rams-horn snail, *Marisa cornuarietis* (Gastropoda: Pilidae) in the USA (T.L. Arsuffi, pers. comm.). Therefore, it is highly probable that the direct and indirect feeding effects of *P. clarkii* contributed to the disappearance of these two species, but the deterioration of the water quality and the damages to macrophyte stands by large herbivore grazing and trampling might have been equally important.

We face a similar problem when analysing the role of *P. clarkii* in the decline of autochtonous crayfish populations. As mentioned above, *P. clarkii* also successfully spreads to areas formerly populated by the native crayfish *Austropotamobius pallipes*. However, it remains unclear whether *P. clarkii* displaced the native species by direct competition or whether it invaded these systems after the populations of *A. potamobius* were already decimated by other mechanisms. In Portugal, *P. clarkii* is mainly restricted to the south-central part, where the native species has never been observed and overlap only occurs in the central part of the country (Anastácio and Marques 1995). Furthermore, the requirements with regard to temperature, water quality and substrate of the two species are quite different. *P clarkii* prefers high temperatures, clayey–silty substrates to construct its burrows and is more tolerant to low water quality, whereas *A. potamobius* lives in temperate to cold waters with coarse substrates and is sensitive to low oxygen and high nutrient concentrations (Gil-Sánchez and Alba-Tercedor 2002). At the moment, data on the autecology of *P. clarkii* from habitats formerly inhabited by the native species are lacking. In zones of abiotic niche overlap, biotic interactions should be intensive and competitive exclusion of the native species might occur. Whether there be direct interaction or not: with the red swamp crayfish present, any recovery of *A. pallipes* populations is unlikely, because *P. clarkii* is also a vector of the aphanomicosis, which is detrimental to the native species.

P. clarkii – a new food item for higher trophic levels

Since its introduction in 1974, *P. clarkii* has been readily accepted as a prey item by fish, birds and mammals thus offering a new resource for higher trophic levels. In some areas such as the Lower Guadalquivir Basin, *P. clarkii* has opened new trophic pathways by transferring energy from the formerly underexploited detritus pool to primary and secondary predators.

Three fish species, six bird species and four mammal species commonly include *P. clarkii* in their diet (Table 3). However, the consumption of crayfish differs considerably according to species, season and study.

For the otter, where information from four quantitative studies over more than one season is available (Adrián and Delibes 1987; Beja 1996; Correia 2001; Ruíz-Olmo et al. 2002), the percentage of crayfish in the total amount of food varies between 1.6 and 76.3% with lowest values in winter and highest in summer (Table 3). All three studies coincide in that otter prey upon crayfish according to crayfish density and prefer small and medium sized individuals. Highest densities of *P. clarkii* in the water bodies coincide with the presence of young otters and feeding on them enhances juvenile survival (Ruíz-Olmo et al. 2002). However, the important bottleneck is in winter, when crayfish are not available, and otters have to rely on scarce native prey species (Beja 1996).

The same is true although to a lesser degree for other mammals such as the red fox (*Vulpes vulpes* L.), the common genet (*Genetta genetta* L.) or the Egyptian mongoose (*Herpestes ichneumon* L.) which also prey upon *P. clarkii* (Correia 2001) (Table 3). As in otters, the highest consumption of crayfish is in summer.

All mammal predators feed in an opportunistic manner on crayfish, and none of them selects this prey item. Diversity of prey in mammals decreases when they start feeding on *P. clarkii*, and crayfish are taken as a function of crayfish density (Correia 2001).

P. clarkii is also an important part of the diet of at least six bird species, in particular for most ciconiiform species. In the case of the white stork, night heron or little egret, crayfish can make up to 80% of the diet during summer, when densities of crayfish are high (Table 3). In addition, other bird species such as the black stork (Parkes et al. 2001) or the lesser black-backed gull (Amat and Aguilera 1988) are reported to feed on *P. clarkii*.

Birds, similar to mammals, consume crayfish above the minimum size for maturity but below the mean size for mature adults (Correia 2001). Predation in this size fragment reduces intraspecific competition among crayfish and produces large-sized adults which in turn produce a higher number of offspring (Correia 2001).

Thus, predation by birds and mammals should help in stock renewal and not negatively affect crayfish populations. Therefore, it remains unclear

Table 3. Frequency of occurrence and percentage of diet of *P. clarkii* in the stomachs of vertebrate predators.

Species	% of occurrence	% of diet			Source
		Mean	Maximum	Biomass	
Fish					
Esox lucius	72.5	82.9		72.4	Elvira et al. (1996)
Micropterus salmoides	5.8	0.9		9.9	García-Berthou (2002)
M. salmoides (> 250 mm; summer)				50–100	García-Berthou (2002)
M. salmoides	72.2				Montes et al. (1993)
Anguilla anguilla	66.7				Montes et al. (1993)
Birds					
Gelochelidon nilotica		40.1		70.1	Costa (1984)
Nycticorax nycticorax		70	71		Correia (2001)
Egretta garzetta		52	86		Correia (2001)
Ardea cinerea		21	40		Correia (2001)
Ardea purpurea		30	31.5		Correia (2001)
Ciconia ciconia		67	86		Correia (2001)
Mammals					
Lutra lutra		67	85		Correia (2001)
	80.3				Adrián and Delibes (1987)
		22.7	42.2		Ruiz-Olmo et al. (2002)
Herpestes ichneumon		26	49		Correia, (2001)
	5.6			1.7	Palomares and Delibes (1991a)
Vulpes vulpes		14	27.5		Correia (2001)
		5	10		Correia (2001)
Genetta genetta	0.8			0.1	

whether the reduction in crayfish numbers observed in the past years is due to increased bird predation or to other factors such as reduced hydroperiods induced by droughts.

Amongst fish, eels (*Anguilla anguilla*) are known to be the most important predators of crayfish (Svardson 1972). In the natural marshlands of the Lower Guadalquivir (Spain), the eels considerably reduced their food spectrum after the red swamp crayfish was introduced, (Table 3). Before introduction, it mainly fed on other fish species such as mosquito fish (*Gambusia affinis*) or carp (*Cyprinus carpio*) which occurred in more than 50% of the stomach contents. After introduction, in 1992, only 16.7% of eel stomachs contained other fish species, and the dominant prey item was *P. clarkii* with a 66.7% occurrence (Montes et al. 1993).

As they readily feed on *P. clarkii*, eels were proposed as effective biological control organisms in a Swiss lake (Mueller and Frutiger 2001). However, eels are also efficient predators of fish eggs and fry as well as of amphibians and reptiles, and therefore, their use to control crayfish populations should be considered with caution.

The other two fish species which include *P. clarkii* in their diet – the northern pike and the largemouth bass (*Micropterus salmoides*) – are both introduced exotics. Crayfish became the dominant prey item of all size classes of pike throughout the year in the Spanish lake system of Ruidera. Crayfish substituted the natural prey species which were reduced near to extinction after the introduction of the pike (Elvira et al. 1996). Without *P. clarkii*, the pike population would have become extinct, as the rest of the fish fauna, mainly composed of other introduced species, could not support self-maintaining pike populations in these lakes. Pike prey on crayfish of a similar size than do birds and mammals (7–9 cm total length).

The largemouth bass (Hickley et al. 1994) readily accepted crayfish as a prey item. In the Guadalquivir marshlands of south-western Spain, it was found to feed exclusively on *P. clarkii* (Montes et al. 1993). In the Spanish lake of Banyoles, dominated by an assemblage of exotic fish, larger size classes of this species (>250 mm) feed predominantly on crayfish except in winter (García-Berthou 2002) (Table 3), a situation typical for water bodies with a low fish diversity (García-Berthou 2002). A similar scenario was described by Hickley et al. (1994) in Lake Naivasha (Kenya), which is also a lake characterised by its low richness in native fish species.

Other species such as perch are known to prey on *P. clarkii*. Perch are able to efficiently reduce densities of *P. clarkii* in mesocosm experiments (Neveu 2001). However, the quantitative impact of the predation of this species on *P. clarkii* populations is not known.

Impact of *P. clarkii* on ecosystem energetics

Besides the impact on structural components of the invaded communities described above, the presence of crayfish might alter to a large degree the pattern of energy flow, especially in systems where detritivores are rare and which are dominated by autotrophs as in temporary freshwater marshes. In such systems, crayfish put the detritus energy pool directly at the disposal of higher trophic levels. This greatly shortens the energy pathways and simplifies their structure (Figure 2). Without crayfish, macrophytes and the associated periphyton are the dominant primary producers in freshwater marshlands from which only a small part of the energy is transmitted to herbivores. Most of the energy is lost to the detritus pool which accumulates high amounts of organic matter. Detritivores, mainly macroinvertebrates (oligochaetes, chironomids) and meiofauna (nematodes, ostracods) are supposed to use only a small fraction of the deposited material. The detritus food chain gains in importance only during drought and refilling of the system in early summer and late autumn, when macrophytes are absent. These systems are characterised by a high diversitiy of herbivores and consist of a minimum of four levels of consumers. Due to the large number of trophic levels and losses of energy to the detritus pool, the energy transferred to top predators such as birds and mammals is comparatively low (Figure 2).

After crayfish introduction, much of the detritus is consumed by this species (Gutiérrez-Yurrita 1997), and the energy gained is directly transferred to the top predator level (fish, birds and mammals). The consequence is a reduction in the number of trophic levels, a decreased importance of macrophytes, herbivores and primary

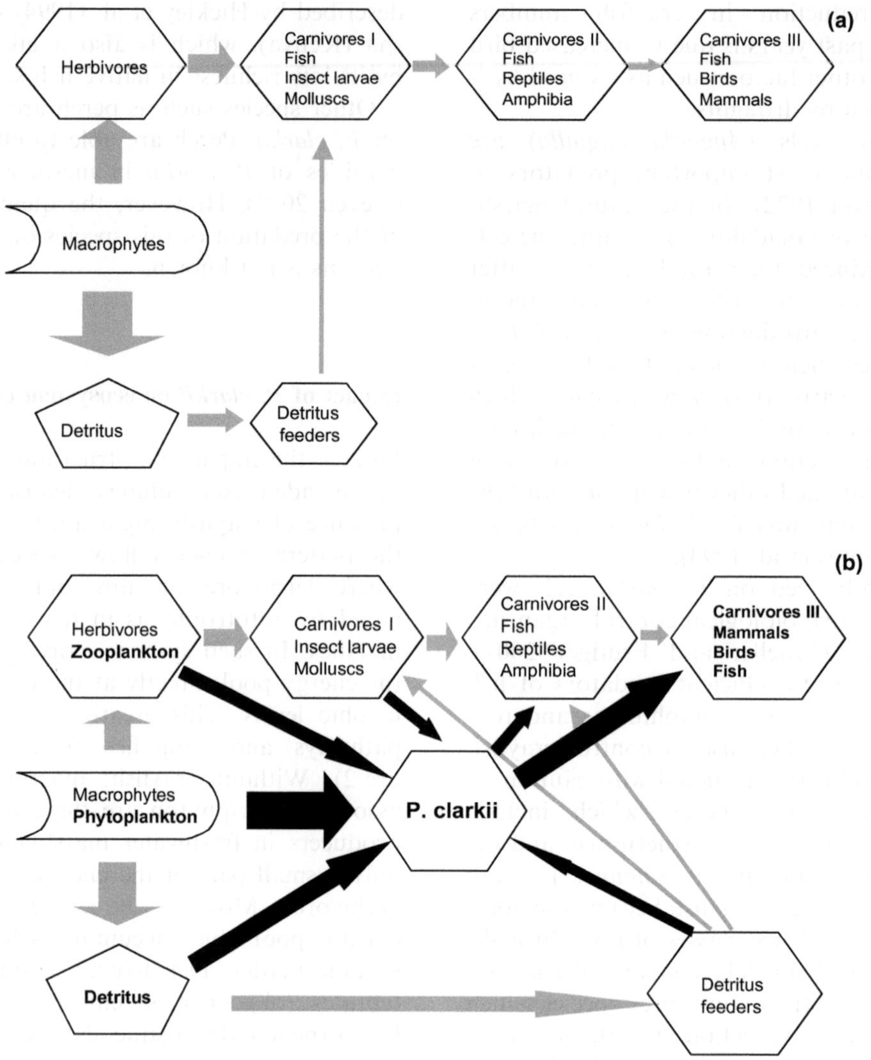

Figure 2. Simplified representation of the energy flow in waterbodies of a freshwater marshland: (a) before the introduction of crayfish and (b) after the introduction of crayfish.

carnivores, but more energy is available for vertebrate predators.

In summary, alien crayfish change both the structure and the functional links of the trophic web in wetlands by opening new resource pathways (detritus food chain), reducing the number of trophic levels, and thus providing more energy to the highest trophic levels.

P. clarkii as a vector of diseases

Introduced crayfish are vectors of several diseases for native crayfishes thus contributing to their decline. One of the most widespread diseases is the 'Crayfish plague', produced by *Aphanomyces astaci* (Schikora), a parasitic saprolegniaceous fungus especially adapted to live in the cuticule of freshwater crayfish (Unestam 1972). This disease has devasted many native European crayfish populations since the 1890s, and the problem became more acute through the massive introductions of American crayfish during the 1960s and 1970s (Persson and Söderhäll 1983; Diéguez-Uribeondo et al. 1995). In Europe, three North American species of crayfish have been shown to carry the infectious fungus in their cuticule: *Pacifastacus leniusculus* (Unestam 1972; Persson and Söderhäll 1983), *Orconectes limosus* (Vey et al. 1983) and

Procambarus clarkii (Diéguez-Uribeondo and Söderhäll 1993). Recent studies using RAPD-PCR have demonstrated the existence of species-specific strains of this fungus. The strain isolated from *P. clarkii* was shown to be the most temperature tolerant (Huang et al. 1994; Diéguez-Uribeondo et al. 1995). The introductions of alien species such as *P. clarkii* also introduced a new *A. astaci* strain with a different genotype and unknown levels of virulence adapted to warm waters (Diéguez-Uribeondo et al. 1995). Recent genetic studies have linked *P. leniusculus* to many recent plague outbreaks in Great Britain (Lilley et al. 1997), Sweden, Finland, Germany and Spain (Diéguez-Uribeondo et al. 1997; Diéguez-Uribeondo 1998).

Other diseases for native species carried by introduced crayfish such as *P. leniusculus*, are the Psorospermiasis, produced by *Psorospermium haeckeli* (Hilgendorf) (Cerenius and Söderhäll 1992; Gydemo 1992; Henttonen et al. 1997), protists which have their phylogenetic roots near the animal–fungal divergence (Ragan et al. 1996).

A question still open for debate is the role of *P. clarkii* in transmitting diseases to humans. An outbreak of tularemia, normally transmitted by small rodents and caused by the bacterium *Francisella tularensis*, in a contaminated stream in central Spain was recently related to *P. clarkii* as a mechanical transmitter (Anda et al. 2001).

P. clarkii – a transmitter of heavy metal contamination

Crayfish have frequently been considered as biological indicators of heavy metal pollution in aquatic environments (Rincón-León et al. 1988). There have been numerous studies on the accumulation of heavy metals in crayfish living in polluted environments (Evans and Edgerton 2002). Most field studies involved chemical analysis of the metal content of crayfish tissues and provided little information on the pathology of heavy metal exposure (Dickson et al. 1979; Finerty et al. 1990; King et al. 1999; MacFarlane et al. 2000; Rowe et al. 2001). There are also many laboratory studies that provide data on the toxicity of metals to freshwater crayfish, the concentrations of metals causing mortality and the pathological effects arising from heavy metal exposure (Bagatto and Alikhan 1987; Naqvi and Flagge 1990; Naqvi et al. 1990; Naqvi and Howell 1993; Reddy et al. 1994; Maranhao et al. 1995; Anderson et al. 1997a, b; Bollinger et al. 1997; Naqvi et al. 1998; Antón et al. 2000). Little attention has been paid to the sublethal pathology of such exposures and how pathological changes could influence the survival of crayfish living in polluted water systems or contaminated culture systems. These kinds of studies is needed for the implementation of adequate restoration and management plans for contaminated areas such as the Guadiamar river basin, which was affected by a toxic spill of approximately 5 Hm^3 of untreated acid fresh water with a high content of metals (especially zinc, copper, cadmium, lead, iron and arsenic) in April 1998 during an accident in Aznalcóllar mine (southwestern Spain). Crayfish captured in this area, have higher heavy metal contents in their tissues compared to those captured outside the contaminated area (Figure 3). They can transfer contaminants to their consumers through bioaccumulation processes (e.g. heavy metals or pesticides enrichment in organs and tissues) (Otero et al. 2003). Other well documented examples of bioaccumulation of heavy metals by red swamp crayfish in Mediterranean wetlands are the studies performed in the rice fields of Albufera Lake in Valencia (eastern Spain) by Díaz-Mayans et al. (1986) and Pastor et al. (1988). These rice fields are surrounded by waters, which received for the last four decades high loads of sewage and toxic industrial residues including heavy metals and pesticides.

Indeed, crayfish are able to effectively regulate the concentration of heavy metals in their tissues (Rainbow and White 1989) and to remove some contaminants from their internal organs and muscles depending on their physiological needs. This is achieved through excretion (faeces) and/or storage in the hepatopancreas – considered the organ of metal storage and detoxification (Alikhan et al. 1990; Anderson et al. 1997a, b; Naqvi et al. 1998) – gills and exoesqueleton (Anderson and Brower 1978; Naqvi et al. 1990; Wright et al. 1991). Consequently, their predators absorb the contaminants immobilised in these crayfish tissues when they ingest them. Measurements of accumulation of heavy metals in waterfowl and other wetland birds living and feeding in the toxic spill area showed that Zn, Cu and As from the spill have entered the food

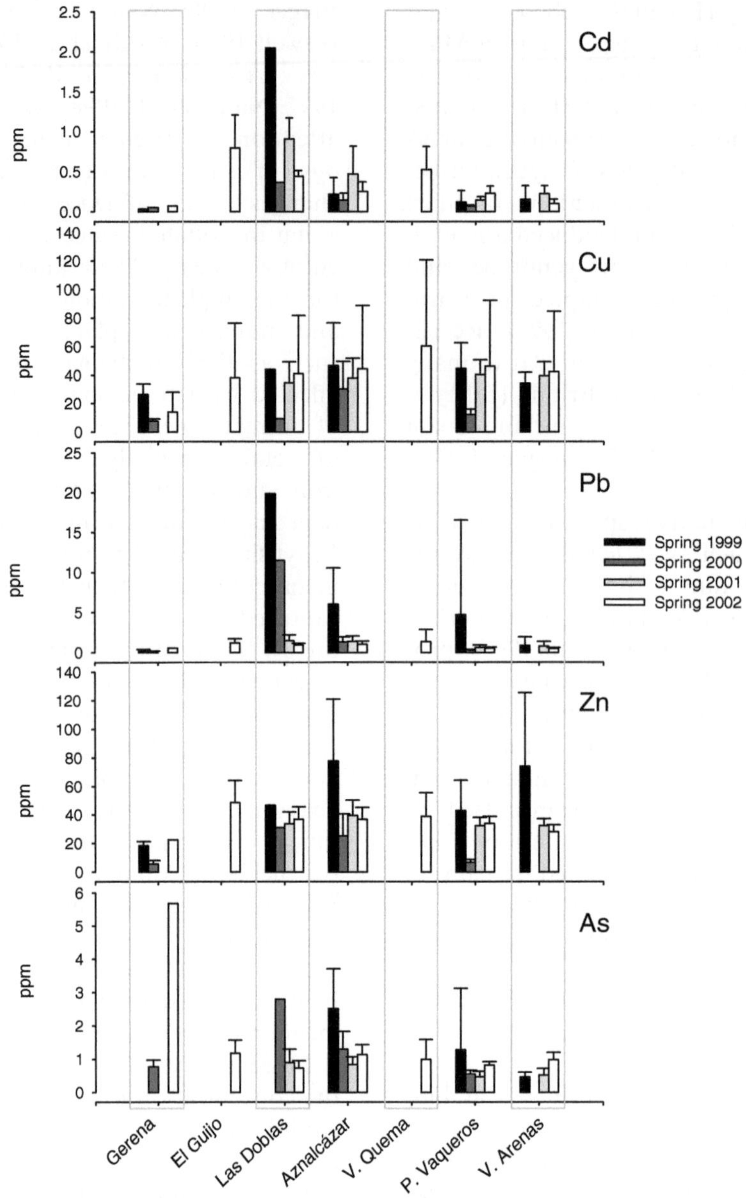

Figure 3. P. clarkii as a vector of contamination for higher trophic levels. Temporal evolution of heavy metal concentrations in crayfish from different sites in the Guadiamar Guadalquivir Basin, southwestern Spain). Bars: mean values; whiskers: standard deviations. Units are in ppm of fresh weight (from Otero et al. 2003).

chain and can be detected in some bird species, such as white stork (*Ciconia ciconia*), spoonbill (*Platalea leucorodia*), or grey heron (*Ardea cinerea*). All these species are fish and crayfish predators (Benito et al. 1999; Hernández et al. 1999). Further studies quantifying the extent of vertebrate contamination through crayfish ingestion are urgently needed.

Impacts derived from the commercial exploitation of *P. clarkii*

In contrast to the significant increase in the attention devoted to the impact that crayfish have on invaded habitats (see above), much less attention is paid to the environmental impact derived from the economic activity which is promoted by the presence of dense crayfish populations or to the socio-economic benefits related to crayfish exploitation. Both kinds of interactions have to be properly considered in any integrative approach to crayfish management.

One important issue to bear in mind is the nature of crayfish exploitation. Extensive crayfish production is not common and is restricted to some areas in the USA and, on a limited basis, in Spain, France, Italy and Zambia (Huner 2002). The most widespread method is the direct use of wild stocks of *P. clarkii* grown in ricefields, irrigation systems, natural marshlands, reservoirs and river deltas. The main crayfish (*P. clarkii*) producer is China, a country that exceeds the production of the USA with 70,000 Tm/year. Spain has developed a much smaller industry (2000–3000 Tm/p year) but of great regional importance (Figure 4a).

Commercial crayfish exploitation in Spain is mainly concentrated in the southwest, in the Lower Guadalquivir Basin. There, *P. clarkii* stocks were intentionally introduced in the early 1970s and immediately developed dense populations within the ricefields. Nowadays, the red swamp crayfish is distributed over almost the whole country with a significant effect on most ecosystems it inhabits. However, exploitation in most places is but recreational with almost no incidence in local economy. Although in these areas the impact of crayfishing should be low, the role of man as a vector of transportation is of major concern. Some areas important for amphibians or fish reproduction can be severely endangered with a single inoculation of just a few animals.

In the area where crayfish exploitation has developed into a growing industry, environmental impact derived from this activity is mainly caused by fishermen during their fishing activity. This impact refers to

(1) physical alteration of the habitat produced by the continuous roaming of the fishermen –

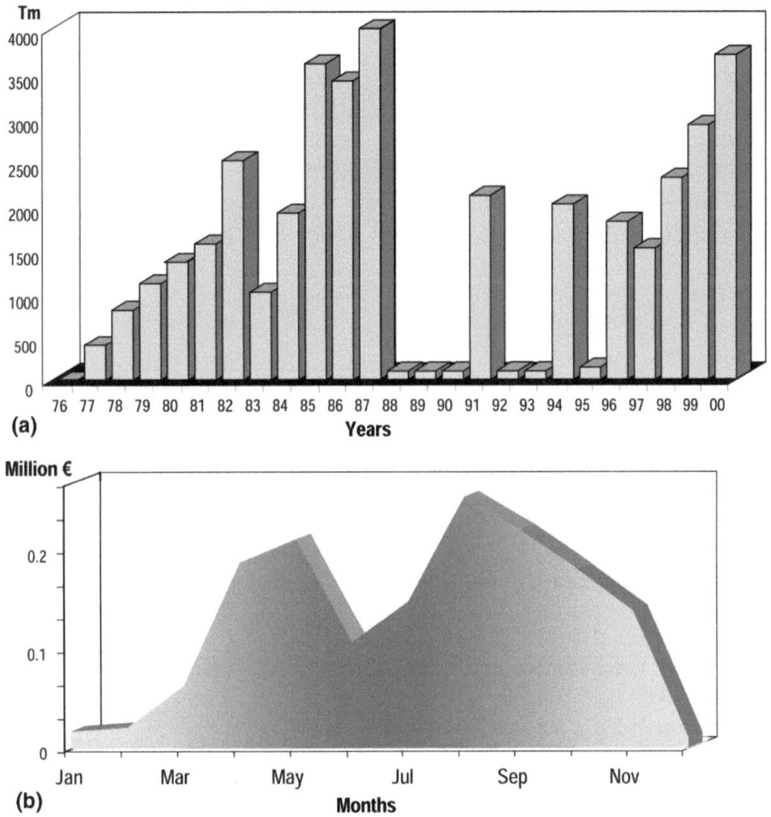

Figure 4. (a) Crayfish commercial captures in Lower Guadalquivir marshlands and (b) commercial value of captured crayfish in 1999.

from 100 to 300 traps/fishermen can be installed simultaneously over large areas; and
(2) the capture of non-target organisms within the crayfish traps.

Fishing activities in natural habitats might severely affect not only habitat structure but also many organisms' reproductive activities. Crayfish exploitation is highly seasonal (Figure 4b) with maxima in late spring and late summer. That first period clearly overlaps with the nesting period of many birds in the area. Intense wandering of people is likely to interfere with reproduction, although no quantitative data are available up to now.

The second kind of impact has been evaluated several times, and results have shown to be relevant for management policies. The traps traditionally used for crayfish were modified eel traps. This kind of trap is not selective for crayfish and, when baited, attracts many different kinds of organisms. Early studies performed in the area (Coronado 1982; Molina and Cadenas 1983; Molina 1984; Domínguez 1987; Asensio 1989) demonstrated the large impact of these traps on birds, amphibians and reptiles. This led to regulation of fishing activities in the area, now strictly forbidden during nesting periods. The impact level on native communities lowered significantly wherever fishing activity was forbidden during the breeding season (Figure 5). Still, a sensible number of non-target organisms die every year in crayfish traps. Some turtle species are of special concern. Most of the victims of this 'collateral damage', however, belong to non-endangered, highly abundant species (Figure 5). Far from being ideal, management of fishing activities – including timing, trap design, and selected locations – can severely reduce the negative impact of crayfishing.

It has to be reminded that crayfish exploitation supports, at least partially, the economy of many families in a poorly developed area and that socio-economic aspects have to be integrated if any management policy is to be developed in order to minimise the negative environmental impact of this alien species.

Conclusions

In the 30 years of invasion history, *P. clarkii* changed the structure and functioning of the invaded ecosystems where it readily occupied a central position in the food webs.

Its success as an invader is mainly due to its adaptation to the main characteristics of Mediterranean wetlands: the frequent periods of drought.

The impact caused by *P. clarkii* affects both lower and higher trophic levels, including grazing

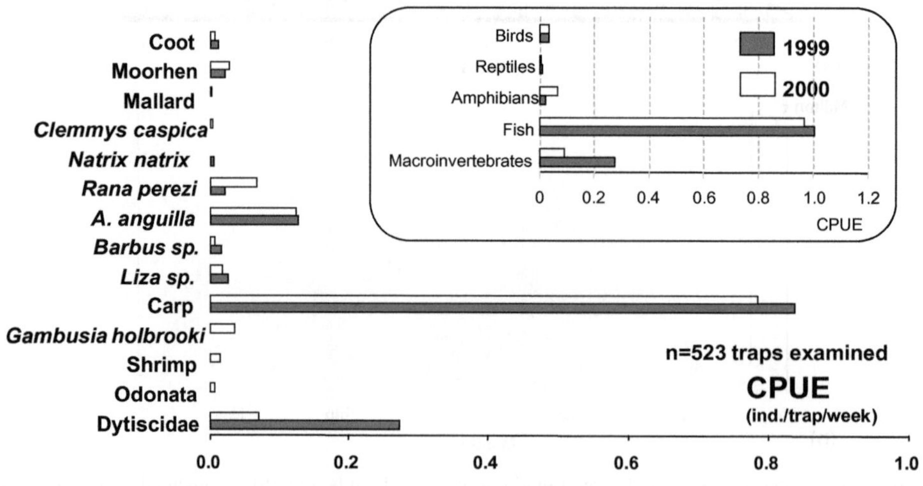

Figure 5. Organisms captured by crayfish traps.

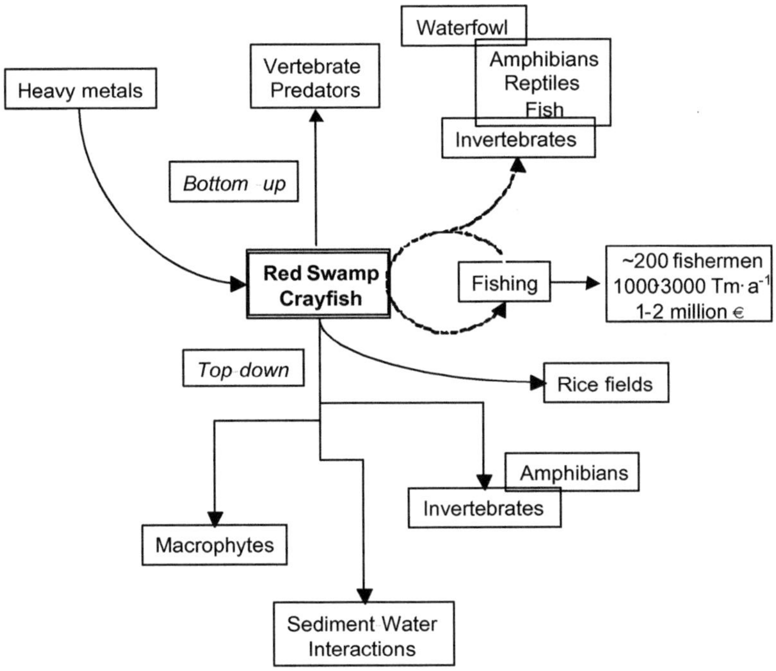

Figure 6. Scheme representing the manyfold impacts of introduced crayfish in Mediterranean wetlands.

on macrophytes, predation on macroinvertebrates and its role as an important food source for numerous vertebrate species (Figure 6).

From a socio-economic point of view, crayfish represent, at least in some areas, an additional, although temporary source of income but cause, on the other hand, serious damage on rice field infrastructure and crayfishing negatively affects vertebrate and invertebrate species.

The numerous and tight links between the invader and the native communities render a successful eradication unlikely. Therefore, control measures to minimise the negative impact should be established. They should include, where possible, a return to natural hydroperiods (salinity limits the distribution of *P. clarkii*) and the implantation of fishing plans with the aim to change the age and size structure of the population. Favouring low density populations dominated by large-sized adults with lower metabolic demands will minimise the impacts caused by this crayfish species. This measurements should be accompanied by a strict control of large herbivore grazing such as cattle and flamingos, water quality and anthropogenic disturbances. All these measures should also favour the recovery of autochthonous crayfish populations.

Prevention is as necessary as control to avoid a further widespread of the species to presently unaffected areas, although most of the suitable habitats seem to be already occupied. We urgently need to protect the remaining areas from being invaded by *P. clarkii* in order to preserve them as ecological reference sites.

Acknowledgements

We are especially grateful to Miguel Angel Bravo, Marina Otero, Yolanda Díaz and José María Martínez for their valuable help with the field work and to Lee Wallace for correcting the English. This work was financed by the 'Consejería de Medio Ambiente, Junta de Andalucía' – Project title: Evaluación del recurso, ordenación pesquera y cultivo del cangrejo rojo (*Procambarus clarkii*) en el Bajo Guadalquivir.

Appendix 1

Effect	Impact on	Habitat type	Location	Source	Comments
Bioturbator	*Tilapia zilli*	Lake	Naivasha (Kenia)	Lowery and Mendes (1977)	Only comments, destroys nesting ground of bottom living *Tilapia*
Bioturbator	Nutrient release	Stream	Degebe stream, Alentejo (Portugal)	Bernardo and Ilhéu (1994)	
Bioturbator	Nutrient release	Ponds	Tablas de Daimiel (Spain)	Angeler et al. (2001)	*P. clarkii* enhances nutrient release from sediment and increases primary production
Bioturbator	Sediment resuspension	Lake	Lago Chozas, León Province (Spain)	Rodríguez et al. (2002)	*P. clarkii* enhances nutrient release from sediment and increases primary production
Competition	Prawns (*Macrobrachium rosenbergii*), Channel catfish (*Ictalurus punctatus*)	Aquaculture ponds	Louisiana (USA)	Huner et al. (1983)	Competition favours large prawns, no effect on catfish fingerlings
Competition	*Austropotamobius pallipes*	Streams and rivers	Province of Granada (Spain)	Gil-Sánchez and Alba-Tercedor (2002)	*P. clarkii* causes regression of *A. pallipes*; cites factors limiting the distribution of *P. clarkii* (altitude, temperature, nutrients, substrate); similar results as in studies of Diéguez-Irubeondo et al. (1997) in Navarra and Bolea-Berné (1995) in Aragon somewhat contradictory as densities of *P. clarkii* are far below those of *A. pallipes* ('not adapted to environment')
Herbivory	Nymphae species	Lake	Naivasha (Kenia)	Lowery and Mendes (1977)	No data, disappearance coincides with introduction of *P. clarkii*
Herbivory	*Elodea*	Laboratory crayfish cultures	Louisiana (USA)	Wiernicki (1984)	Fresh and as detritus; 15-day detritus best assimilated
Herbivory	Macrophytes (*Potamogeton pectinatus*)	Pond	Fremont, California (USA)	Feminella and Resh (1989)	With crayfish exclusion: six-fold macrophyte biomass and higher *Anopheles* densities
Herbivory	Macrophytes (*Potamogeton pectinatus*)	Experimental mesocosms in marsh	Coyote Hills Marsh, California (USA)	Feminella and Resh (1987)	Enclosure experiments showed a strong positive relationship between crayfish density and macrophyte clearance
Herbivory	Macrophytes	Laboratory crayfish cultures	North Carolina (USA)	Bolser et al. (1998)	Preference experiments, defense mechanisms of plants decide about preferences
Herbivory	Macrophytes	Laboratory crayfish cultures		Cronin (1998)	Preference experiments among nine species of submersed, floating, emergent, and shoreline macrophytes
Herbivory	Macrophytes	Laboratory crayfish cultures	North Carolina (USA)	Cronin et al. (2002)	Preference experiments among 14 species of freshwater macrophytes (including macroscopic algae)
Herbivory	Macrophytes, detritus	Stream	Degebe stream (Portugal)	Ilhéu and Bernardo (1995)	Preference experiments: prefer high organic matter contents, high protein and low fibre, prefer 'fresh' detritus to macrophytes

Predator	*Corbicula* sp. (Asiatic clam)	Laboratory ponds	Oklahoma (USA)	Covich et al. (1980)	*P. clarkii* feeds on damaged *Corbicula*, cites also predation of *Orconectes limosus* on *Dreissena* in Poland (Piesik, 1974), not as efficient as *O. limosus* due to different shape of prey and that it is not used to prey
Predator	Amphibia (eggs and tadpoles)	Freshwater marsh	Doñana National Park, Arroyo Rocina, Lucio Bolín (Spain)	Delibes and Adrián (1987)	
Predator	Two high pelletised diets and a third one formulated with fishmeal,	Laboratory crayfish cultures	Baja California, Mexico	Cordero and Voltolina (1990)	*P. clarkii* responds better to two of the diets at lower temperatures (20 °C)
Predator	*Rana perezi* (eggs and tadpoles) and other macroinvertebrates	Freshwater marsh, rice fields	Ebro Delta (Spain)	Ibañez et al. (2000)	
Prey	Black bass (*Micropterus salmoides*)	Lake	Naivasha (Kenia)	Lowery and Mendes (1977)	Pers. comm.: 70% of the food of black bass
Prey	Dragonfly (*Anax junius*)	laboratory crayfish cultures	Louisiana (USA)	Witzig et al. (1986)	Max 1. 16 crayfish/day; depends on temperature and crayfish size. Do not coincide temporally in the field
Prey	Gull-billed tern (*Gelochelidon nilotica*)	Freshwater and saline marsh, rice fields	Guadalquivir basin (Spain)	Costa (1984)	*P. clarkii* represents 70% of the biomass of the diet
Prey	Otter (*Lutra lutra*)	Freshwater marsh	Doñana National Park: Arroyo Rocina, Lucio Bolín (Spain)	Adrian and Delibes (1987)	*P. clarkii* occurred in 80% of the faeces from Arroyo de la Rocina
Prey	Otter (*Lutra lutra* L.) Mentions also (without data): Red fox (*Vulpes vulpes*), Polecat (*Mustela putorius*), Common genet (*Genetta genetta*), Egyptian mongoose (*Herpestes ichneumon*), Badger (*Meles meles*). Birds, Ciconiiformes: Night heron (*Nycticorax nycticorax* L.), Grey heron (*Ardea cinerea* L.), Purple heron (*Ardea purpurea* L.), Little egret (*Egretta grazetta* L.), White stork (*Ciconia ciconia* L.) Raptors: Black kite (*Milvus migrans*), Tawny owl (*Strix aluco*)	Freshwater marsh	Doñana National Park	Delibes and Adrián (1987)	Adaptation to new food within five years. Most important prey after five years. Increasing importance of insects is an indirect effect of crayfish hunting. Emphasises on the negative consequences of a possible eradication for otters but need of control

Appendix 1. Continued.

Effect	Impact on	Habitat type	Location	Source	Comments
Prey	Black-backed Gull (*Larus fuscus*)	Freshwater marsh, rice fields, streams, channels, ponds	Guadalquivir basin (Spain)	Amat and Aguilera (1988)	All attacks performed by the Gulls were done on birds carrying *P. clarkii* in their mouth
Prey	Egyptian mongoose (*Herpestes ichneumon*), Common genet (*Genetta genetta*).	Freshwater marsh	Doñana National Park (Spain)	Palomares and Delibes (1991a, b)	Appear in 5.6% of the mongooses and in 0.8% of the genets
Prey	Otter (*Lutra lutra*)	Torgal stream	Alentejo (Portugal)	Beja (1996)	Crayfish and eels were particularly important in the diet from April to October. For the rest of the year, crayfish accounted for < 10% of the monthly energetic intake, and cyprinids and toads were the most important prey.
Prey	Pike (*Esox lucius*)	Lakes	Ruidera lakes (Spain)	Elvira et al. (1996)	Exotic–exotic interaction; *P. clarkii* the dominant prey item under all conditions (frequency of occurrence 72. 55%, rel. importance 70%)
Prey	Waterfowls	Freshwater marsh, rice fields	Ebro Delta (Spain)	Ibañez et al. (2000)	
Prey	Mammals, Carnivora: Red fox (*Vulpes vulpes L.*), Otter (*Lutra lutra L.*), Common genet (*Genetta genetta L.*), Egyptian mongoose (*Herpestes ichneumon L.*). Birds, Ciconiiformes: Night heron (*Nycticorax nycticorax* L.), Grey heron (*Ardea cinerea* L.), Purple heron (*Ardea purpurea* L.), Little egret (*Egretta grazetta* L.), White stork (*Ciconia ciconia* L.)	Freshwater marsh, ponds, reservoirs, rice fields	Tejo river basin (Portugal)	Correia (2001)	Most predators prey on *P. clarkii* more in spring summer and autumn than in winter; *L. lutra* (67% of diet); mongoose (26%), genet (5%). Follows abundance pattern of *P. clarkii*. Prey size below mean size at reproduction, mature females are in burrows, feeding on sub-adults leads to larger size at maturity, reduces intra-specific competition.
Prey	Black stork (*Ciconia nigra*)	Rice fields and drainage channels	Las Cabezas de San Juan, Guadalquivir Basin (Spain)	Parkes et al. (2001)	*P. clarkii* is the major contribution to the diet of black storks in this area
Prey	Otter (*Lutra lutra*)	Rivers	Pyrenees, Ebro basin (Spain)	Ruiz-Olmo et al. (2002)	Timing of breeding with availability of *P. clarkii*, 46% of diet
Secondary effects	Amphibia, reptiles	Freshwater marsh	Doñana National Park, Arroyo Rocina, Lucio Bolín (Spain)	Delibes and Adrián (1987)	No direct results but mentions egg and tadpole predation and secondary effects by getting trapped in fishing nets
Secondary effects	Marbled teal (*Marmaronetta angustirostris*)	Rice fields, channels, freshwater marsh	Guadalquivir basin (Spain)	Aguayo and Ayala (2002)	Endangered avian species; gets trapped in crayfish nets

Secondary effects	Native crayfish species	Laboratory crayfish cultures	North Carolina (USA)	Antonelli et al. (1999)	Juvenile *P. clarkii* and adults prefer different shelter, *Orconectes* the same, advantage for *P. clarkii*
Secondary effects	Native crayfish species	Commercial cultures	Laboratory	Figler, Cheverton and Blank (1999)	Competition for shelter might expose native crayfish species to increased predation
Secondary effects	White stork (*Ciconia ciconia*)	Freshwater marsh, rice fields	Doñana National Park, Guadalquivir basin (Spain)	Negro and Garrido-Fernández (2000)	Astaxanthin: red colour of skin and tarsi – potential negative effects on mating behaviour
Secondary effects	Native crayfish species (*Austropotamobius pallipes*)	Streams and rivers	Zamora province (Spain)	Palacios and Rodríguez (2002)	*P. clarkii* acts as an afanomicosis vector infecting native crayfish
Secondary effects	Native crayfish species (*Austropotamobius pallipes*)	Streams and rivers	Granada province (Spain)	Gil and Alba-Tercedor (1998)	*P. clarkii* acts as an afanomicosis vector infecting native crayfish
Secondary effects	Native crayfish species (*Austropotamobius pallipes*)	Streams and rivers	Valencia province (Spain)	Monzó et al. (2001)	*P. clarkii* acts as an afanomicosis vector infecting native crayfish
Secondary effects	Native crayfish species (*Austropotamobius pallipes*)	Streams and rivers	Spain	Diéguez and Rueda (1994)	*P. clarkii* and *Pacifastacus leniusculus* act as afanomicosis vectors infecting native crayfish
Secondary effects	Waterfowls	Lake	Lago Chozas, León Province (Spain)	Rodríguez et al. (2002)	The density of waterfowls has diminished because of the lack of food consequence of the activity of *P. clarkii* that has destroyed submersed macrophytes
Economic losses	Destruction of fishnets and damage on fish yield in nets	Lake	Naivasha (Kenia)	Lowery and Mendes (1977)	30% damage on fish yield
Economic losses	Rice field structures	Rice fields, freshwater marsh	Guadalquivir basin (Spain)	Algarín (1980)	Damages on rice field installations – no data
Economic losses	Rice field structures	Rice fields, freshwater marsh	Guadalquivir basin (Spain)	Gaudé (1984)	Damage to dams and irrigation structures due to deeper and more complicated burrows in semi-permanent water bodies
Economic losses	Rice field structures	Rice fields, freshwater marsh	Guadalquivir basin (Spain)	Serres, González and Parrondo (1985)	Damages on rice field installations – no data
Economic losses	Rice field structures	Rice fields, freshwater marsh, reservoirs	Portugal	Correia and Ferreira (1995)	Damage to dams and irrigation structures due to deeper and more complicated burrows in semi-permanent water bodies
Economic losses	Rice field structures	Rrice fields, irrigation channels	Guadalquivir basin (Spain)	Cano and Ocete (1997)	Damage to dams and irrigation structures due to deeper and more complicated burrows in semi-permanent water bodies
Economic losses	Rice field structures	Rice fields, irrigation channels	Guadalquivir basin (Spain)	Aguayo and Ayala (2002)	Damage to dams and irrigation structures due to deeper and more complicated burrows in semi-permanent water bodies

References

Abrahamsson SAA (1966) Dynamics of an isolated population of the crayfish, *Astacus astacus* Linne. Oikos 17: 96–107

Adao H and Marques JC (1993) Population biology of the red swamp crayfish *Procambarus clarkii* (Girard 1852) in southern Portugal. Crustaceana 63(3): 336–345

Adrian MI and Delibes M (1987) Food habits of the otter (*Lutra lutra*) in two habitats of the Doñana National Park, SW Spain. Journal of Zoology, London 212: 399–406

Aguayo M and Ayala J (2002) Siguen muriendo cercetas pardillas en nasas para pescar cangrejo rojo. Quercus 199: 48–49

Alcorlo P, Geiger W and Otero M (in press) Feeding preferences and food selection of the red swamp crayfish (*Procambarus clarkii*) in habitats differing in food item diversity. Crustaceana

Alderman DJ (1996) Geographical spread of bacterial and fungal disease of crustaceans. Rev. Sci. Tech. O I E (Off. Int. Epizoot.) 5(2): 603–632

Alderman DJ and Polglase JL (1988) Pathogens, parasites and commensals. In: Holdich DM and Lowery RS (eds) Freshwater Crayfish: Biology, Management and Exploitation, pp 167–212. Croom Helm, London

Algarín S (1980) Problemática y perspectiva de la introducción de los cangrejos americanos en las marismas del Bajo Guadalquivir. In: El cangrejo rojo de las marismas, pp 25–31, Consejería de Agricultura y Pesca, Junta de Andalucía, Spain

Alikhan MA, Bagatto G and Zia S (1990) The crayfish as a 'biological indicator' of aquatic contamination by heavy metales. Water Research 24(9): 1069–1076

Amat JA and Aguilera E (1988) Robo de alimento a aves acuáticas por gaviotas sombrias (*Larus fuscus*). Ardeola 35(2): 275–278

Anastácio PM and Marques JC (1995) Population Biology and production of the red swamp crayfish *Procambarus clarkii* (Girard) in the lower Mondego river Valley, Portugal. Journal of Crustacean Biology 15(1): 156–168

Anda P, Segura J, Díaz JM, Escudero R, García FJ, López MC, Selleck RE, Jiménez MR, Sánchez LP and Martínez JF (2001) Waterborne outbreak of tularemia associated with crayfish fishing. Emerging Infectious Diseases 7(3): 575–582

Anderson MB, Preslan JE, Jolibois L, Bollinger JE and George WJ (1997a) Bioaccumulation of lead nitrate in red swamp crayfish (*Procambarus clarkii*). Journal of Hazardous Materials 54: 15–26

Anderson MB, Reddy P, Preslan JE, Fingerman M, Bollinger J, Jolibois L, Maheshwarudu G and George WJ (1997b) Metal accumulation in crayfish, *Procambarus clarkii*, exposed to a petroleum-contaminated Bayou in Louisiana. Ecotoxicology and Environmental Safety 37: 267–272

Anderson RV and Brower JE (1978) Patterns of trace metal accumulation in crayfish populations. Bulletin of Environmental Contamination and Toxicology 20: 120–127

Angeler DG, Sánchez-Carrillo S, García G and Alvarez-Cobelas M (2001) The influence of *Procambarus clarkii* (Cambaridae, Decapoda) on water quality and sediment characteristics in a Spanish floodplain wetland. Hydrobiologia 464: 89–98

Antón A, Serrano T, Angulo E, Ferrero G and Rallo A (2000) The use of two species of crayfish as environmental quality sentinels: the relationship between heavy metal content, cell and tissue biomarkers and physico-chemical characteristics of the environment. The Science of the Total Environment 247: 239–251

Antonelli J, Steele C and Skinner C (1999) Cover-seeking behavior and shelter use by juvenile and adult crayfish, *Procambarus clarkii*: potential importance in species invasion. Journal of Crustacean Biology 19(2): 293–300

Arrignon JC, Gérard P, Krier A and Laurent PJ (1999) The situation in Belgium, France and Luxembourg. In: Gherardi F and Holdich DM (eds) Crayfish in Europe as Alien Species. How to Make the Best of a Bad Situation?, pp 129–140. A. A. Balkema, Rotterdam, The Netherlands

Asensio JM (1989) Impacto de la captura del cangrejo rojo sobre otras poblaciones animales del Brazo del Este. Informe Inédito. Junta de Andalucía

Avault JW, Romaire RP and Miltner RM (1983) Feeds and forages for red swamp crawfish, *P. clarkii*: 15 years research at Louisiana State University reviewed. Freshwater Crayfish V: 362–369

Bagatto G and Alikhan MA (1987) Copper, cadmium and nickel accumulation in crayfish populations near copper–nickel smelters ar Sudbury, Ontario, Canada. Bulletin of Environmental Contamination and Toxicology 38: 540–545

Baker HG (1974) The evolution of weeds. Annual Review of Ecology and Systematics 5: 1–24

Beja PR (1996) An analysis of otter *Lutra lutra* predation on introduced American crayfish *Procambarus clarkii* in Iberian streams. Journal of Applied Ecology 33: 1156–1170

Benito V, Devesa V, Muñoz O, Suñer MA, Montoro RBR, Hiraldo F, Ferrer M, Frenández M and González MJ (1999) Trace elements in blood collected from birds feeding in the area around Doñana National Park affected by the toxic spill from the Aznalcóllar mine. The Science or the Total Environment 242: 1309–1323

Bernardo JM and Ilhéu M (1994) Red swamp crayfish (*Procambarus clarkii*): Contribution to material cycling. Verhandlungen der Internationalen Vereinigung für Limnologie 25: 2447–2449

Bollinger JE, Bundy K, Anderson MB, Millet L, Preslan JE, Jolibois L, Chen HL, Kamath B and George WJ (1997) Bioaccumulation of chromium in Red Swamp crayfish (*Procambarus clarkii*). Journal of Hazardous Materials 54: 1–13

Cano E and Ocete ME (1997) Population biology of red swamp crayfish, *Procambarus clarkii* (Girard 1852) in the Guadalquivir river marshes, Spain. Crustaceana 70(5): 553–561

Carpenter SR and Lodge DM 1986. Effects of submersed macrophytes on ecosystem processes. Aquatic Botany 26: 341–370

Cerenius L and Söderhäll K (1992) Crayfish diseases and crayfish as vectors for important diseases. Finnish Fisheries Research 14: 125–133

Chambers PA and Hanson JM (1990) The impact of the crayfish *Orconectes virilis* on aquatic macrophytes. Freshwater Biology 24: 81–91

Chambers PA, Hanson JM, Burke JM and Prepas EE (1990) The impact of the crayfish *Orconectes virilis* on aquatic macrophytes. Freshwater Biology 24: 81–91

Chidester FE (1908) Note on the daily life and food of *Cambarus bretoni*. American Naturalist 42: 710–716

Costanza R, d'Arge R, de Groot R, Farber S, Grasso M, Hannon B, Limburg K, Naeem S, O'Neill RV, Paruelo J, Raskin RG, Sutton P and van den Belt M (1997) The value of the world's ecosystem services and natural capital. Nature 387: 253–260

Cordero B and Voltolina D (1990) Short-term evaluation of three pelletized diets for the red swamp crayfish *Procambarus clarkii* (Girard). Anales del Instituto de Ciencias del Mar y Limnología 17(1). Retrieved from http://biblioweb.dgsca.unam.mx/cienciasdelmar/instituto/1990-1/articulo362.html on 21 november 2002

Coronado R (1982) Resumen-informe sobre el cangrejo rojo en la campaña de 1983. ICONA. Ministerio de Agricultura

Correia AM (1992) A note on the occurrence of white-eyed red swamp crayfish, *Procambarus clarkii* (Decapoda: Cambaridae) in Portugal. Arquivos Do Museu Bocage II(11): 257–261

Correia AM (2001) Seasonal and interspecific evaluation of predation by mammals and birds on the introduced red swamp crayfish *Procambarus clarkii* (Crustacea, Cambaridae) in a freshwater marsh (Portugal). Journal of Zoology, London 255: 533–541

Correia AM and Costa AC (1994) Introduction of the red swamp crayfish *Procambarus clarkii* (Crustacea, Decapoda) in Sao Miguel, Azores, Portugal. Arquipélago 12(A): 67–73

Correia AM and Ferreira O (1995) Burrowing behaviour of the introduced red swamp crayfish *Procambarus clarkii* (Decapoda, Cambaridae) in Portugal. Journal of Crustacean Biology 15(2): 248–257

Covich P (1977) How do crayfish respond to plants and Mollusca as alternate food resources? Freshwater Crayfish III: 165–179

Covich AP, Dye LL and Mattice JS (1980) Crayfish predation on *Corbicula* under laboratory conditions. The American Midland Naturalist 105(1): 181–188

Creed RP (1994) Direct and indirect effects of crayfish grazing in a stream community. Ecology 75(7): 2091–2103

Cronin G (1998) Influence of macrophyte structure, nutritive value, and chemistry on the feeding choices of a generalist crayfish. In: Jeppesen E, Sondegaard MA, Sondegaard MO and Christoffersen K (eds) The Structuring Role of Submerged Macrophytes in Lakes, pp 307–317. Springer-Verlag, New York

Cronin G, Lodge DM, Hay ME, Miller M, Hill AM, Horvath T, Bolser RC, Lindquist N and Wahl M (2002) Crayfish feeding preferences for freshwater macrophytes: the influence of plant structure and chemistry. Journal of Crustacean Biology 22(4): 708–718

Dean JL (1969) Biology of the crayfish, *Orconectes causeyi*, and its use for control of aquatic weeds in trout lakes. US Department of the Interior, Fish and Wildlife Service, Bureau of Sprots Fisheries and Wildlife. Technical Report 24

Delibes M and Adrián I (1987). Effects of crayfish introduction on otter *Lutra lutra* in the Doñana National Park, SW Spain. Biological Conservation 42: 153–159

Di Castri F (1990) On invading species and invaded ecosystems: the interplay of historical chance and biological necessity. In: Di Castri F, Hansen AJ and Debussche M (eds) Biological Invasions in Europe and the Mediterranean Basin, pp 3–16. Kluwer Academic Publishers, Dordrecht, The Netherlands

Díaz-Mayans J, Hernández F, Medina J, Del Ramo J and Torreblanca A (1986) Cadmium accumulation in the crayfish, *Procambarus clarkii*, using graphite furnace atomic absorption spectroscopy. Bulletin of Environmental Contamination and Toxicology 37: 722–729

Dickson GW, Briese LA and Giesy JP (1979) Tissue metal concentrations in two crayfish species cohabiting a Tennessee cave stream. Oecologia 44: 8–12

Diéguez-Uribeondo J (1998) El cangrejo de río: distribución, patología, inmunología y ecología. AquaTIC 3. Retrieved from http://www.revistaaquatic.com on 13 January 2003

Diéguez-Uribeondo J and Söderhäll K (1993) *Procambarus clarkii* as a vector for the crayfish plague fungus *Aphanomyces astaci* Schikora. Aquaculture and Fisheries Management 24: 761–765

Diéguez-Uribeondo J, Huang T, Cerenius L and Söderhäll K (1995) Physiological adaptation of an *Aphanomyces astaci* strain isolated from the freshwater crayfish *Procambarus clarkii*. Mycological Research 9(5): 574–578

Diéguez-Uribeondo J, Temiño C and Muzquiz J (1997) The crayfish plague fungus (*Aphanomyces astaci*) in Spain. Bulletin Français de la Pêche y Pisciculture 347: 753–763

Domínguez, L (1987) Impacto de la pesca del cangrejo rojo americano (Procambarus clarkii, Girard) en el Parque Nacional de Doñana durante la temporada 1987. ICONA. Ministerios de Agricultura, Pesca y Alimentación

Duarte C, Montes C, Agustí S, Martino P, Bernués M and Kalff J (1990) Biomasa de macrófitos acuáticos en la marisma del Parque Nacional de Doñana (SW de España): importancia y factores ambientales que controlan su distribución. Limnetica 6: 1–12

Elvira B, Nicola GG and Almodóvar A (1996) Pike and red swamp crayfish: a new case on predator–prey relationship between aliens in central Spain. Journal of Fish Biology 48: 437–446

Enríquez S, García-Murillo P, Montes C and Amat JA (1987) Macrófitos acuáticos de la laguna costera de El Portil (Huelva). IV Congreso Español De Limnología. Sevilla

Evans LH and Edgerton BF (2002) Pathogens, parasites and commensals. In: Holdich DM (ed) Biology of Freshwater Crayfish, pp 377–464. Blackwell Science, Oxford

Feminella JW and Resh VH (1986) Effects of crayfish grazing on mosquito habitat at Coyote Hills Marsh. In: Proceedings of the Fifty-Fourth Annual Conference of the California Mosquito and Vector Control Association, USA, pp 101–104

Feminella JW and Resh VH (1989) Submerged macrophytes and grazing crayfish: an experimental study of herbivory in California freshwater marsh. Holartic Ecology 12: 1–8

Figler MH, Cheverton HM and Blank GS (1999) Shelter competition in juvenile red swamp crayfish (*Procambarus clarkii*): the influences of sex differences, relative sizes, and prior residence. Aquaculture 178: 63–75

Finerty M, Madden J, Feagley S and Grodner R (1990) Effect of environs and seasonality on metal residues in tissues of wild and pond raised crayfish in Southern Louisiana. Archives of Environmental Contamination and Toxicology 19: 94–100

Flint RW and Goldman CR (1975) The effect of a benthic grazer on the primary productivity of the littoral zone of Lake Tahoe. Limnology and Oceanography 20: 935–944

García-Berthou E (2002) Ontogenetic diet shifts and interrupted piscivory in the introduced Largemouth Bass (*Micropterus salmoides*). Internationale Revue für Hydrobiologie 87(4): 353–363

Gaude AP (1984) Ecology and production of Louisiana red swamp crayfish *Procambarus clarkii* in Southern Spain. Freshwater Crayfish 6: 111–130

Gherardi F (2002) Behaviour. In: Holdich DM (ed) Biology of Freshwater Crayfish, pp 258–290. Blackwell Science, Oxford

Gherardi F, Baldaccini GN, Barbaresi S, Ercolini P, De Luise G, Mazzoni D and Maurizio M (1999) The situation in Italy. In: Gherardi F and Holdich DM (eds) Crayfish in Europe as Alien Species. How to Make the Best of a Bad Situation?, pp 107–128. A. A. Balkema, Rotterdam, The Netherlands

Gil JM and Alba-Tercedor J (1998) El cangrejo de río autóctono en la provincia de Granada. Quercus 144: 14–15

Gil-Sánchez JM and Alba-Tercedor J (2002) Ecology of the native and introduced crayfishes *Austropotamobius pallipes* and *Procambarus clarkii* in southern Spain and implications for conservation of the native species. Biological Conservation 105: 75–80

Goddard JS (1988) Food and feeding. In: Holdich DM and Lowery RS (eds) Freshwater Crayfish, Management and Exploitation, pp 145–166. Croom Helm, London

Grillas P, García-Murillo P, Geertz-Hansen N, Marbá C, Montes C, Duarte CM, Tan-Ham L and Grossman A (1993) Submerged macrophyte seed bank in a Mediterranean temporary marsh: abundance and relationship with established vegetation. Oecologia 94: 1–6

Gutiérrez-Yurrita PJ (1997) El papel ecológico del Cangrejo Rojo, *Procambarus clarkii* en los ecosistemas acuáticos del Parque Nacional de Doñana. Una perspectiva ecofisiológica y bioenergética. PhD Thesis. Dpto de Ecología, Universidad Autónoma de Madrid

Gutiérrez-Yurrita PJ, Sancho G, Bravo MA, Baltanás A and Montes C (1998) Diet of the red swamp crayfish *Procambarus clarkii* in natural ecosystems of the Doñana National Park temporary fresh-water marsh (Spain). Journal of Crustacean Biology 18(1): 120–127

Gutiérrez-Yurrita PJ and Martínez JM (2002) Analyse ecologique de l'impact ambiant de la population du ecrevisse rouge (*Procambarus clarkii)*, a Tenerife, Îles Canaries, Espagne, et ses formes de minorizer. L'Astaciculteur de France 71: 2–12

Gydemo R (1992) Crayfish diseases and management – the need for knowledge. Finnish Fisheries Research 14: 119–124

Habsburgo-Lorena AS (1983) Socioeconomic aspects of the crawfish industry in Spain. Freshwater Crayfish 5: 552–554

Hanson JM, Chambers PA and Prepas EE (1990) Selective foraging by the crayfish *Orconectes virilis* and its impact on macroinvertebrates. Freshwater Biology 24: 69–80

Henttonen P, Huner JV, Rata P and Lindqvist OV (1997) A comparison of the known life forms os *Psorospermium* spp. in freshwater crayfish (Arthropoda, Decapoda) with emphasis on *Astacus astacus* L. (Astacidae) and *Procambarus clarkii* (Girard) (Camabaridae). Aquaculture 149(1–2): 15–30

Hernández LM, Gómara B, Fernández M, Jiménez B, González MJ, Baos R, Hiraldo F, Ferrer M, Benito V, Suñer MA, Devesa V, Muñoz O and Montoro R (1999) Accumulation of heavy metals and As in wetland birds in the area around Doñana National Park affected by the Aznalcóllar toxice spill. The Science of the Total Environment 242: 293–308

Hickley P, North R, Muchiri SM and Harper DM (1994) The diet of largemouth bass, *Micropterus salmoides*, in Lake Naivasha, Kenya. Journal of Fisheries Biology 44: 607–619

Hill AM and Lodge DM (1999) Evaluating competition and predation as mechanisms of crayfish species replacements. Ecological Applications 9: 678–690

Hobbs HH Jr (1988) Crayfish distribution, adaptive radiation and evolution. In: Holdich DM (ed) Freshwater Crayfish: Biology, Management, and Exploitation, pp 52–82. Croom Helm, London

Hobbs HH Jr (1993) Trophic Relationships of North American Freshwater Crayfish and Shrimps. Hobbs HH III. Contributions in Biology and Geology, Vol 85. Milwaukee Public Museum, 110 pp

Hobbs HH Jr, Jass JP and Huner JV (1989) A review of global crayfish introductions with particular emphasis on two North American species (Decapoda, Cambaridae). Crustaceana 56(3): 299–316

Holdich DM (1997) Negative effects of established crayfish introductions. In: Gherardi F (ed) The Introduction of Alien Species of Crayfish in Europe, pp 9–11. University of Florence, Florence

Holdich DM (1999a) The negative effects of established crayfish introductions. In: Gherardi F and Holdich DM (eds) Crayfish in Europe as Alien Species. How to Make the Best of a Bad Situation?, pp 31–47. A.A. Balkema, Rotterdam, The Netherlands

Holdich DM (1999b) The introduction of alien crayfish into Britain for commercial purposes an own goal? In: The Biodiversity Crisis and Crustacea: Proceedings of the Fourth International Crustacean Congress, Amsterdam, The Netherlands, 20–24 July 1998, pp 85–97. A.A. Balkema, Rotterdam, The Netherlands

Holdich DM (2002) Background and functional morphology. In: Holdich DM (ed) Biology of Freshwater Crayfish, pp 3–29. Blackwell Science, Oxford

Huang T, Cerenius L and Söderhäll K (1994) Analysis of genetic diversity in the crayfish plague fungus, *Aphanomyces astaci*, by random amplification of polymorphic DNA. Aquaculture 126: 1–10

Huner JV (1981) Information about the biology and culture of the red crawfish, *Procambarus clarkii* (Girard 1852) (Decapoda, Cambaridae) for fisheries managers in Latin America. Anales del Instituto de Ciencias del Mar y Limnología 8(1): 43–50

Huner JV (2002) *Procambarus*. In: Holdich DH (ed) Biology of Freshwater Crayfish, pp 541–574. Blackwell Science, Oxford

Huner JV, Miltner M, Avault JW and Bean RA (1983) Interactions of freshwater prawns, channel catfish fingerlings, and crayfish in earthen ponds. The Progressive Fish Culturalist 45(1): 36–40

Huryn AD and Wallace JB (1987) Production and litter processing by crayfish in an Appalachian mountain stream. Freshwater Biology 18: 277–286

Ibañez C, Canicio A, Curcó A and Riera X (2000) El proyecto Life del Delta del Ebro (SEO/Birdlife). Boletín SEDHUMED 16: 4–6

Ilhéu M and Bernardo JM (1993) Experimental evaluation of food preference of red swamp crayfish, *Procambarus clarkii*: vegetal versus animal. Freshwater Crayfish 9: 359–364

Ilhéu M and Bernardo JM (1995) Trophic ecology of red swamp crayfish *Procambarus clarkii* (Girard) – preferences and digestibility of plant foods. Freshwater Crayfish 10: 132–139

King FH (1883) The food of the crayfish. American Naturalist 17: 980–981

King HM, Baldwin DS, Rees GN and McDonald S (1999) Apparent bioaccumulation of Mn derived from paper-mill effluent by the freshwater crayfish Cherax destructor – the role of Mn oxidising bacteria. The Science of the Total Environment 226: 261–267

Lilley JH, Cerenius L and Söderhäll K (1997) RAPD evidence for the origin of crayfish plague outbreaks in Britain. Aquaculture 157: 181–185

Lodge DM and Hill AM (1994) Factors governing species composition, population size, and productivity of coolwater crayfishes. Nordic Journal of Freshwater Research 69: 111–136

Lodge DM and Lorman JG (1987) Reductions in submersed macrophyte biomass and species richness by the crayfish *Orconectes rusticus*. Canadian Journal of Fisheries and Aquatic Sciences 44: 591–597

Lodge DM, Taylor CA, Holdich DM and Skurdal J (2000a) Nonindigenous crayfishes threaten North American freshwater biodiversity: lessons from Europe. Fisheries 25(8): 7–20

Lodge DM, Taylor CA, Holdich DM and Skurdal J (2000b) Reducing impacts of exotic crayfish. Introductions: new policies needed. Fisheries 25(8): 21–23

Lorman JG and Magnuson JJ (1978) The role of crayfishes in aquatic ecosystems. Fisheries 3: 8–10

Lowery RS and Mendes AJ (1977) *Procambarus clarkii* in the lake Naivasha, Kenya, and its effects on established and potential fisheries. Aquaculture 11: 111–121

MacFarlane GR, Booth DJ and Brown KR (2000) The semaphore crab, *Heloecius cordiformis*: bio-indication potential for heavy metals in estuarine systems. Aquatic Toxicology 50: 153–166

Maranhao P, Marques JC and Madeira V (1995) Copper concentrations in soft tissues of the red swamp crayfish *Procambarus clarkii* (Girard 1852), after exposure to a range of dissolved copper concentrations. Freshwater Crayfish 10: 282–286

Marçal-Correia A (2003) Food choice by the introduced crayfish *Procambarus clarkii*. Annales Zoologici Fennici 40: 517–528

Molina F (1984) La pesca del Cangrejo Rojo Americano y su influencia en el entorno del Parque Nacional de Doñana. Revista de Estudios Andaluces 3: 151–160

Molina F and Cadenas R (1983) Impacto de la pesca del cangrejo rojo americano (*Procambarus clarkii*) en los ecosistemas marismeños del Parque Nacional de Doñana durante la campaña de 1983. ICONA, Ministerio de Agricultura y Pesca, Madrid, Spain

Momot WT (1967) Population dynamics and productivity of the crayfish *Orconectes virilis*, in a Marl Lake. American Midland Naturalist 78: 55–81

Momot WT (1995) Redefining the role of crayfish in aquatic ecosystems. Reviews in Fisheries Science 3(1): 33–63

Montes C and Bernués M (1991) Incidencia del flamenco rosa (*Phoenicopterus ruber roseus*) en el funcionamiento de los ecosistemas acuáticos de la marisma del Parque Nacional de Doñana (SW España). In: Reunión técnica sobre la situación y problemática del flamenco rosa (*Phoenicopterus ruber roseus*) en el Mediterráneo occidental y África noroccidental, pp 103–110. Junta de Andalucía, Agencia de Medio Ambiente, Spain

Montes C, Bravo-Utrera MA, Baltanás A, Duarte C and Gutiérrez-Yurrita PJ (1993) Bases ecológicas para la gestión del Cangrejo Rojo de las Marismas en el Parque Nacional de Doñana. ICONA, Ministerio de Agricultura y Pesca, Madrid, Spain

Monzó J, Sancho V and Galindo J (2001) Estado y distribución actual del cangrejo de río autóctono (*Austropotamobius pallipes*) en la Comunidad Valenciana. AquaTIC 12. Retrieved from http://www.revistaaquatic.com/aquatic/art.asp?t = h&c = 100 on 12 December 2002

Mueller R and Frutiger A (2001) Effects of intensive trapping and fish predation on an (unwanted) population of *Procambarus clarkii*. In: Abstracts of Annual Meeting North American Benthological Society, LaCrosse, Wisconsin, 3–8 June 2001

Naqvi SM and Flagge CT (1990) Chronic effects of arsenic on american red crayfish, *Procambarus clarkii*, exposed to monosodium methanearsonate (MSMA) herbicide. Bulletin of Environmental Contamination and Toxicology 45: 101–106

Naqvi SM and Howell RD (1993) Cadmium and lead uptake by red swamp crayfish (*Procambarus clarkii*) of Louisiana. Bulletin of Environmental Contamination and Toxicology 51: 296–302

Naqvi SM, Flagge CT and Hawkins RL (1990) Arsenic uptake and depuration by red crayfish, *Procambarus clarkii*, exposed to various concentrations of monosodium methanearsonate (MSMA) herbicide. Bulletin of Environmental Contamination and Toxicology 45: 94–100

Naqvi SM, Devalraju NH and Naqvi NH (1998) Copper bioaccumulation and depuration by red swamp crayfish, *Procambarus clarkii*. Bulletin of Environmental Contamination and Toxicology 61: 65–71

Negro JJ and Garrido-Fernández J (2000) Astaxanthin is the major carotenoid in tissues of white storks (*Ciconia ciconia*) feeding on introduced crayfish (*Procambarus clarkii*). Comparative Biochemistry and Physiology 126 (Part B): 347–352

Neveu A (2001) Can resident carnivorous fish slow down introduced alien crayfish spread? Efficacity of 3 fish species versus 2 crayfish species in experimental design. Bulletin Français de la Pêche et Pisciculture 361: 683–704

Nyström P and Strand JA (1996) Grazing by a native and an exotic crayfish on aquatic macrophytes. Freshwater Biology 36: 673–682

Nyström P, Brönmark C and Graneli W (1996) Patterns in benthic food webs: a role for omnivorous crayfish? Freshwater Biology 36: 631–646

Olsen TM, Lodge DM, Capelli GM and Moulihan RJ (1991) Mechanisms of impact of an introduced crayfish (*Orconectes rusticus*) on littoral congeners, snails and macrophytes. Canadian Journal of Fisheries and Aquatic Sciences 48: 1853–1861

Otero M, Díaz Y, Martínez JM, Baltanás A, Montoro R and Montes C (2003) Efectos del vertido minero de Aznalcóllar sobre las poblaciones de cangrejo rojo americano (*Procambarus clarkii*) del río Guadiamar y Entremuros. In: Corredor Verde del Guadiamar (eds) Ciencia y restauración del río Guadiamar. Resultados del programa de investigación del Corredor Verde del Guadiamar 1998–2002, pp 126–137. Consejería de Medio Ambiente, Junta de Andalucía, Spain

Palacios J and Rodríguez M (2002) La provincia de Zamora se queda sin cangrejos de río autóctonos. Quercus 192: 50–51

Palomares F and Delibes M (1991a) Dieta del meloncillo, *Herpestes ichneumon*, en el Coto del Rey (Norte del Parque Nacional de Doñana, SO de España). Doñana Acta Vertebrata 18(2): 187–194

Palomares F and Delibes M (1991b) Alimentación del meloncillo *Herpestes ichneumon* y de la gineta *Genetta genetta* en la Reserva Biológica de Doñana, S.O. de la Península Ibérica. Doñana Acta Vertebrata 18(1): 5–20

Parkes C, Torés-Ruiz A and Torés-Sánchez A (2001) Población invernante de Cigüeña negra (*Ciconia nigra*) en los arrozales junto al río Guadalquivir (1998–2001). Retrieved from http://www.terra.es/personal4/aletor on 14 February 2003

Pastor A, Medina J, Del Ramo J, Torreblanca A, Díaz-Mayans J and Fernández F (1988) Determination of lead in treated crayfish *Procambarus clarkii*: accumulation in diferent tissues. Bulletin of Environmental Contamination and Toxicology 41: 412–418

Penn GH (1943) A study of the life history of the Louisiana red crayfish, *Procambarus clarkii* Girard. Ecology 24(1): 1–18

Perry WL, Lodge DM and Lamberti GA (1997) Impact of crayfish predation on exotic zebra mussels and native invertebrates in a lake-outlet stream. Canadian Journal of Fisheries and Aquatic Sciences 54: 120–125

Persson M and Söderhäll K (1983) *Pacifastacus leniusculus* Dana and its resistance to the parasitic fungus *Aphanomyces astaci* Schikora. Freshwater Crayfish 5: 292–298

Power ME and Tilman D (1996) Challenges in the quest for keystones. Bioscience 46(8): 609–620

Prins R (1968) Comparative ecology of the crayfishes, *Orconectes rusticus* and *Cambarus tenebrosus* in Doe Run, Meade County, Kentucky. Internationale Revue der Gesamten Hydrobiologie 53: 667–714

Ragan MA, Goggin CL, Cawthorn JR, Cerenius L, Jamieson AV, Plourde SM, Rand TG and Söderhäll K (1996) A novel clade of protistan parasites near the animal-fungus divergence. Proceedings of the National Academy of Sciences, USA 93(21): 11907–11912

Rainbow P and White SL (1989) Comparative strategies of heavy metal accumulation by crustaceans: zinc, Cu and cadmium in decapod, an amphipod and a barnacle. Hydrobiologia 174: 245–262

Ramos MA and Pereira MG (1981) Um novo Astacidae para a fauna portuguesa: *Procambarus clarkii* (Girard 1852). Boletim do Instituto de Investigação das Pescas (Lisboa) 6: 37–47

Reddy PS, Devi M, Sarojini R and Nagabhushanam R (1994) Cadmium chloride induced hyperglycemia in the red swamp crayfish, *Procambarus clarkii*: possible role of crustacean hyperglycemic hormone. Comparative Biochemistry and Physiology 107C(1): 57–61

Rickett PJ (1974) Trophic relationships involving crayfish of the genus *Orconectes* in experimental ponds. The Progressive Fish Culturalists 36: 207–211

Rincón-León F, Zurera-Cosano G and Pozo-Lora R (1988) Lead and cadmium concentrations in red crayfish (*Procambarus clarkii* G.) in the Guadalquivir River marshes (Spain). Archives of Environmental Contamination and Toxicology 17: 251–256

Rodríguez CF, Bécares E and Fernández-Aláez M (2002) El cangrejo rojo americano *Procambarus clarkii*, como mecanismo de pérdida de la fase clara en la Laguna de Chozas (León). In: Abstracts of XI Congreso de la Asociación Española de Limnología y III Congreso Ibérico de Limnología, Madrid, Spain, 17–22 June 2002

Rosenthal SK, Lodge DM, Muohi W, Ochieng P, Chen T, Mkoji G and Mavuti K (2001) Louisiana crayfish in Kenyan ponds: non-target effects of a potential biocontrol agent. In: Abstracts of Annual Meeting North American Benthological Society, LaCrosse, Wisconsin, 3–8 June 2001

Rowe CL, Hopkins WA, Zehnder C and Congdon JD (2001) Metabolic costs incurred by crayfish (*Procambarus acutus*) in a trace element-polluted habitat: further evidence of similar responses among diverse taxonomic groups. Comparative Biochemistry and Physiology Part C 129: 275–283

Ruiz-Olmo J, Olmo-Vidal JM, Manás S and Batet A (2002) The influence of resource seasonality on the breeding pattern of the Eurasian otter (*Lutra lutra*) in Mediterranean habitats. Canadian Journal of Zoology 80: 2178–2189

Saiki MK and Tash JC (1979) Use of cover and dispersal by crayfish to reduce predation by large mouth bass. In: Johnsonn DL and Stein RA (eds) Response of Fish to Habitat Structure in Standing Water, pp 44–48. North Central Division American Fisheries Society Spatial Publication

Sakai AK, Allendorf FW, Holt JS, Lodge DM, Molofsky J, With KA, Baughman S, Cabin JC, Cohen JE, Ellstrand NC, McCauley DEOP, Parker IM, Thompson JN and Weller SG (2001) The population biology of invasive species. Annual Review of Ecology and Systematics 32: 305–332

Scholtz G, Braband A, Tolley L, Reimann A, Mittmann B, Lukhaup C, Steuerwald F and Vogt G (2002) Parthenogenesis in an outsider crayfish. Nature 421: 806

Serres JM, González C and Parrondo JL (1985) Problemática del cangrejo de río en España. ICONA, Ministerio de Agricultura y Pesca, Madrid, Spain

Shea K and Chesson P (2002) Community ecology theory as a framework for biological invasions. Trends in Ecology and Evolution 17(4): 170–176

Smith VJ and Söderhäll K (1986) Crayfish Pathobiology: an overview. Freshwater Crayfish 6: 199–211

Stucki TP (1997) Three american crayfish species in Switzerland. Freshwater Crayfish 11: 130–133

Stucki TP and Staub E (1999) Distribution of crayfish species and legislation concerning crayfish in Switzerland. In: Gherardi F and Holdich DM (eds) Crayfish in Europe as Alien Species. How to Make the Best of a Bad Situation?, pp 141–147. A. A. Balkema, Rotterdam, The Netherlands

Svardson G (1972) The predatory impact of the eel, *Anguilla anguilla*, on populations of the crayfish *Astacus astacus*. Reprints of the Institute of Freshwater Research, Drottningholm 52: 149–191

Tack PI (1941) The life history and ecology of the crayfish, *Cambarus immunis* Hagen. American Midland Naturalist 25: 420–466

Taugbol T and Skurdal J (1993) Noble Crayfish in Norway: legislation and yield. Freshwater Crayfish 9: 134–143

Taugbol T, Skurdal J and Hastein T (1993) Crayfish plague and management strategies in Norway. Biological Conservation 63: 75–82

Unestam T (1972) On the host range and origin of the crayfish plague fungus. Reprints of the Institute of Freshwater Research, Drottningholm 52: 192–198

Vey A, Söderhäll K and Ajaxon A (1983) Susceptibility of *Orconectes limosus* Raff. to the crayfish plague: *Aphanomyces astaci* Schikora. Freshwater Crayfish 5: 284–291

Webster JR and Patten BC (1979) Effects of watershed perturbation on stream potassium and calcium dynamics. Ecological Monographs. 49(1): 51–72

Westman K and Westman P (1992) Present status of crayfish management in Europe. Finnish Fisheries Research 14: 1–22

Whiteledge GW and Rabeni CF (1997) Energy sources and ecological role of crayfishes in an Ozark stream: insights from stable isotopes and gut analysis. Canadian Journal of Fishery and Aquatic Sciences 54: 2555–2563

Wiernicki C (1984) Assimilation efficiency by *Procambarus clarkii* fed Elodea (*Egera densa*) and its products of decomposition. Aquaculture 36(3): 203–215

Witzig JF, Huner JV and Avault JWJr (1986) Predation by dragonfly Naiads *Anax janius* on young crawfish *Procambarus clarkii*. Journal of the World Aquaculture Society 17(4): 58–63

Wright DA, Welbourn PM and Martin AVM (1991) Inorganic and organic mercury uptake and loss by the crayfish *Orconectes propinquus*. Water, Air, and Soil Pollution 56: 697–707

Biological Invasions (2005) 7: 75–85

Loss of diversity and degradation of wetlands as a result of introducing exotic crayfish

C.F. Rodríguez[1,*], E. Bécares[2], M. Fernández-Aláez[2] & C. Fernández-Aláez[2]
[1]*Instituto de Medio Ambiente, Universidad de León, La Serna 56, 24071 León, Spain;* [2]*Departamento de Ecología, Facultad de Biología, Universidad de León, Campus de Vegazana, 24071 León, Spain;*
**Author for correspondence (e-mail: degcrv@unileon.es; fax: +34-987-291501)*

Received 4 June 2003; accepted in revised form 30 March 2004

Key words: alien species, aquatic birds, biodiversity, crayfish, ecosystem, lake, macrophytes, wetland

Abstract

The introduction of the alocthonous Louisiana red swamp crayfish (*Procambarus clarkii*) in Chozas (a small shallow lake situated in León (North-West Spain)) in 1996 switched the clear water conditions that harboured an abundant and a quite high richness of plants, invertebrates, amphibians and birds to a turbid one followed by strong losses in abundance and richness in the aforementioned groups. Crayfish exclusion experiments done in Chozas previous to this work confirmed the role of crayfish herbivorism on macrophyte destruction that had a trophic cascade effect on the wetland ecosystem. Direct and indirect effects of crayfish introduction on Chozas lake communities have been evaluated and compared with previous conditions before 1996 or with other related lakes in which crayfish were no present. Crayfish had a main role in submerged plant destruction and a potential effect on amphibia and macroinvertebrate population decrease. Plant destruction (99% plant coverage reduction) was directly related to invertebrates (71% losses in macroinvertebrate genera), amphibia (83% reductions in species), and waterfowls (52% reduction). Plant-eating birds were negatively affected (75% losses in ducks species); nevertheless, fish and crayfish eating birds increased their presence since the introduction. Introduction of crayfish in shallow plant-dominated lakes in Spain is a main risk for richness maintenance in these endangered ecosystems.

Introduction

Many studies include the fundamental role of aquatic vegetation in maintaining the ecological integrity of wetlands (e.g. Jeppessen et al. 1990; Moss 1990; Scheffer 1990). Submerged macrophytes participate in a series of feedback mechanisms so that various physical and chemical variables remain within the limits appropriate to their development in the presence of aquatic vegetation. This submerged vegetation also supports a complex trophic chain, completely different from and much more diverse than those present in wetlands without vegetation (Carpenter and Lodge 1986).

Shallow lakes can change abruptly from a situation characterised by abundant submerged vegetation and transparent waters to another characterised by the absence of vegetation and very turbid conditions due to the dominance of phytoplankton. These two extreme situations are considered alternative stable states related to the nutrient load received by the lake (Scheffer 1990). As eutrophication increases, the submerged vegetation is capable of controlling the growth of phytoplankton via diverse mechanisms (allelopathy, zooplankton shelter, nutrient competition, etc. (Jeppesen and Sammalkorpi 2002)) up to rather high nutrient concentrations. Below these concentrations, the disappearance of vegetation due to different causes other than nutrient increase produces a rapid change to high turbidity conditions dominated by phytoplankton.

Amongst the most frequently occurring causes of submerged vegetation destruction, we can cite anthropogenic interferences such as direct mechanical destruction, increased internal load of nutrients due to waste from agricultural farms, alteration of the hydrological cycle due to abusive contributions or extractions (Blindow et al. 1998), the arrival of biocides and other chemical compounds (Vandergaag 1992), excess herbivorism by macroinvertebrates, mammals and birds (Lodge et al. 1998) and the recent introduction of exotic species, especially various decapods (Nyström 1999). Rodríguez et al. (2003) have recently shown that herbivorism by alien crayfish can change eutrophic systems of clear waters previously dominated by vegetation to a turbid state.

For some decades, exotic species of freshwater crayfish has been deliberately introduced. Various hydrographic basins in most parts of Europe have been supplied with alien crayfish from North America (Ackefors 1999), mainly *Procambarus clarkii, Pacifastacus leniusculus and Orconectes limosus*, although other species are being commercially exploited. This worsens the situation of autochthonous decapods even more, as the exotic North American species are vectors of the *Aphanomyces astaci* fungus, though resistant to it (Dieguez-Uribeondo and Soderhall 1993).

Given that many of the freshwater decapod species are generalist omnivores (Cronin 1998; Nyström 1999; Parkyn et al. 2001), the effect of introducing them into aquatic systems is not restricted to displacement of the autochthonous ones but also extends to the rest of the trophic chain levels, favouring species removal and destabilising the self-control mechanisms of submerged vegetation (Axelsson et al. 1997; Gutiérrez-Yurrita et al. 1998; Nyström 1999).

Various types of impact have been documented as regards the introduction of alien crayfish, such as competitive removal of autochthonous crayfish (Soderback 1994; Hill and Lodge 1999) and fish species (Guanz and Wiles 1997; Dorn and Mittelbach 1999), the disappearance of amphibians due to predation (Axelsson et al. 1997; Gerardhi et al. in press), declines in macroinvertebrate fauna (Hanson et al. 1990; Lodge et al. 1998; Nyström and Perez 1998; Nyström et al. 2001; Correia 2002) or the destruction of the aquatic plant cover (Feminella and Resh 1989; Olsen et al. 1991; Nyström 1999; Rodríguez et al. 2003).

The red crayfish of Louisiana (*Procambarus clarkii*) is a generalist crayfish first introduced in Spain in 1974 (Gutiérrez-Yurrita and Montes 1999); since then it has spread throughout Europe. Its impact on wetlands has been described in terms of changes in available food and/or shelter and the breeding success of other species (Gutiérrez-Yurrita and Montes 1999) and other direct effects on the aquatic fauna and flora (Cronin 1996; Nyström and Pérez 1998; Angeler et al. 2001; Rodríguez et al. 2003). However, the extent of these effects on the systems and their canalisation through the trophic chains have not been precisely determined (Nyström 1999).

This paper gives evidence of the effects of predation by an exotic crayfish (*P. clarkii*) on the trophic web of Chozas Lake (León, northwestern Spain) as well as other indirect effects related to the destruction of the aquatic plant habitat. The destabilisation of the trophic chains has produced serious losses in biodiversity and, for the first time, the introduction of an exotic macroinvertebrate can be related to the disappearance not only of aquatic fauna (amphibians and macroinvertebrates) but also various species of birds.

Study area

Chozas Lake is a small, shallow mass of water (9 ha; maximum depth 1.8 m) situated in León (northwestern Spain) surrounded by a small area of meadow wetlands which houses an important nesting colony of lapwings. The lake was traditionally used as an irrigation pool for agriculture, and so water-containing walls were built on the south and west sides in the 1950s to increase its depth. Characteristically, the water level fluctuates greatly, both monthly due to the dry summer (it drops to almost 1 m as regards maximum capacity, and the flooded area decreases to about 3 ha during the summer), and from year to year (serious drought in 1993). The inhabitants of the area have traditionally used the lake for marginal uncontrolled hunting (bird and amphibian hunting, fishing).

The first limnological study was made by Fernández-Aláez (1984) as an example of a wetland with abundant and diverse aquatic vegetation. The lake has been regularly monitored since 1994 onwards.

Materials and methods

This study reviews the existing information on the Chozas Lake communities before the introduction of crayfish and compares it with later studies. The plant biomass samplings and species catalogue done by Fernández-Aláez (1984, 1999) prior to the introduction of crayfish was used as the basis for checking changes in flora. Main nutrients (total phosphorous (TP), soluble reactive phosphorous (SRP), nitrates and ammonium) as well as chorophyll *a* levels and other physical parameters (mainly Secchi disc values that measure light penetration through a water column) following APHA-AWWA-WPCF Standard Methods for the Examination of Water and Wastewater (1989) have been taken from the same sources. The statistical significance of nutrients and Chl. *a* levels were tested by means of Student's *t*-test.

The fauna inventories prior to wetland degradation used as a reference are those by Álvarez and Salvador (1984) for amphibian fauna. The data on over-wintering, breeding and passing birds in Chozas Lake are taken from Alegre et al. (1991), Marcos et al. (1995) and data provided in Purroy (1990). Due to the fact that these studies partially overlap as regards various groups of birds, the inventory with the largest record of species and/or specimens was used to establish the number of species affected and the decrease. The inventory of rare and/or threatened species (according to Blanco and González 1992) present in Chozas was taken from the Purroy (1990) report. Current aquatic bird inventories and populations have been determined since 1999 by fortnightly counting using a 60 magnification × 80 mm Kowa telescope and 8 × 20 Zenit binoculars.

The number of crayfish in Chozas was estimated in a previous study (Rodríguez et al. 2003), where a capture–marking–recapture method was used (Krebs 1991), carried out in September 2002. The capture method chosen was that of batteries of small baited traps. The mark used was the same for all the specimens and captures and consisted of a tangential cut in one of the telson parts. The study showed that the number of crayfish in Chozas is around 1 individuals/m^2, with a mean weight of 20 g/specimen. This allows the biomass of red crayfish in Chozas Lake to be calculated as approximately 200 kg/ha. The fish species present along 15 min transects were determined at the same time using an electric fishing technique; then, species present and their numbers as well as the weights and the lengths of individuals were recorded.

Due to the lack of data referring to the macroinvertebrate fauna for Chozas Lake prior to the arrival of *P. clarkii*, the mean values of the number of Genera and Families present in various wetlands around the study area (García-Criado et al. 2004) were used as a reference, differentiating between macroinvertebrates associated with vegetation and benthic ones. The benthic macroinvertebrate samplings in Chozas were carried out using two sampling nets of different pore sizes. Those of macroinvertebrates associated with vegetation were carried out in transects with sampling nets and a Kornijów sampler. For more details on methodology and results, consult (García-Criado et al. 2004).

Results

Changes in physical and chemical variables

From 1984 to 1996, the lake was characterised by the abundant submerged vegetation and great transparency of its waters. Although an increase in the total phosphorus (TP) values from 30 to 60 μg/l was recorded at the beginning of the 1990s, the mean nutrient and chlorophyll values did not vary significantly between 1984 and 1996 (Table 1). After the destruction of the aquatic vegetation in the summer of 1997, the lake switched to the turbid state. The nutrient concentration increased significantly, although some concrete fractions did not change; nevertheless, a significant depletion of nitrates ($P < 0.05$; *t*-test for independent samples) has been noticed. The most important variations correspond to the increase recorded in TP levels. These concentrations rose by 800% (statistically significant; $P < 0.05$) in the first few years after the arrival of the crayfish; in the following years the TP

Table 1. Physical and chemical characterisation of Chozas Lake in periods before and after exotic crayfish introduction, with indication of maximun and minimun values.

	Before 1997	After 1997
Total phosphorous (μg/l)		
Mean	38.8	226.6
Min.	18.801 (SP 94)	23.168 (SP 01)
Max.	69.2 (WI 96)	665.1 (SU 99)
Phosphates (μg/l)		
Mean	8.86	7.9
Min.	0.1 (AU 94)	0.0 (SU)
Max.	37.9 (SP 95)	51.7 (SU 98)
Nitrates (mg/l)		
Mean	0.1	0.0
Min.	0.190 (SP 94)	0:0 (year)
Max.	5.47 (SU 96)	0.11 (AU 99)
Ammonium (mg/l)		
Mean	49.6	13.5
Min.	11.1 (SP 94)	0.0 (SU)
Max.	161.0 (SP 95)	179.2 (AU 99)
Chlorophyll a (mg/l)		
Mean	16.3	68.5
Min.	5.8 (SU 96)	2.5 (SP 01)
Max.	61.6 (WI 96)	161.4 (SU 99)
Secchi (cm light penetration)		
Mean	Bottom	47.6
Min.	No data	19 (SU 99)
Max.	No data	Bottom (SP 01)
Inorg. suspended solids (mg/l)	No data	Max. 22

SP: spring, SU: summer, AU: autumn, WI: winter.

levels stabilised at around 130 μg/l. The water chlorophyll content has increased 100% ($P < 0.05$; *t*-test for independent samples), with mean concentrations of 69 mg/l in summer at present, so the present trophic status in Chozas can be considered hypertrophic (OECD 1982). This, together with the greatly increased resuspension of the sediment due to wind and waves, as well as to crayfish benthic activities has reduced light penetration of the water to a Secchi disc depth of 28 cm.

Submerged aquatic vegetation

Plant cover did not vary significantly in Chozas Lake from 1984 to 1996 with 95% of the bottom kept covered by a varied community of macrophytes (Fernández-Aláez et al. 1999) (Table 2): *Chara globularis, Nitella translucens, Myriophyllum alterniflorum* and *Potamogeton natans* dominated in the deepest areas, *Baldellia ranunculoides, Littorella uniflora, Glyceria fluitans, Juncus heterophyllus, Sparganium erectum* and *Eleocharis palustris* covered the shallowest ones.

Table 2. Floristic list and species coverage in Chozas Lake in summer. Comparison of three years.

Species	% Cover		
	1981	1995	2001
Potamogeton natans	25	20	Absent
Chara globularis	20	25	+
Myriophyllum alterniflorum	15	6	0.5
Littorella uniflora	12	10	1
Nitella translucens	10	12	Absent
Eleocharis palustris	8	12	+
Antinoria agrostidea	4	3	+
Glyceria fluitans	1	2	+
Baldellia ranunculoides	1	1	+
Galium palustre	1	1	+
Juncus heterophyllus	+	2	+
Utricularia australis	+	+	Absent
Ranunculus peltatus	+	+	+
Apium inundatum	+	+	Absent
Scirpus fluitans	+	+	Absent
Mentha pulegium	+	1	+
Total surface covered by macrophytes	97%	95%	<2%

+ = presence.

During the beginning of the summer of 1997, most of the submerged vegetation in the lake was destroyed; the biomass value in summer fell from 800 gDW in 1996 to 70 gDW (a decrease of over 90%), although the different species of macrophytes were affected differently (Fernández-Aláez et al. 2002). Whilst most of the angiosperm stems gradually appeared on the shores over the summer, the dense clumps of Charophytes disappeared completely. Since 1998, no submerged vegetation has appeared in the lake, although there is weak colonisation of the banks at the beginning of spring in areas that dry out in summer, which means that most species are still present but at very low coverage (see Table 2). Currently, the vegetation cover is 2% in early summer, but no macrophyte species can be found within the flooded perimeter as summer advances; in mid-summer, any submerged plant biomass can be recorded.

Macroinvertebrates

Due to the lack of data on macroinvertebrate communities prior to 1997, Table 3 compares the

Table 3. Epiphytic (3a) and benthic (3b) macroinvertebrate groups present in Chozas compared to lakes not colonised by exotic crayfish species (crayfishless). Data from García-Criado et al. (2004).

	Crayfishless	Chozas
Epiphytic macroinvertebrates		
% Hydra	0.3	0
% Oligochaeta	2.1	1
% Hirudinea	0.7	0
% Gastropoda	14	0
% Acari	1.3	0
% Ostracoda	0.8	0.3
% Ephemeroptera	5.4	13.3
% Odonata	55.6	78.6
% Heteroptera	5.4	0
% Lepidoptera	4.8	0
% Trichoptera	3	0
% Diptera	19.5	6.8
% biomass	3.2% biomass in Chozas	
Benthic macroinvertebrates		
% Oligochaeta	8.2	17.6
% Diptera	52.9	82,4
% Others	40.0	0.0
% Biomass	233% higher in Chozas	

The % of biomass corresponds to the ratio: crayfishless lakes biomass/Chozas biomass.

Chozas communities with those of other crayfishless lakes in the area. According to the samplings carried out after crayfish were introduced in Chozas, the current benthic macroinvertebrate fauna (three genera) represents only about 26% of those present in the lakes not invaded by exotic decapods. On the other hand, 72% of frequently appearing genera associated with aquatic vegetation detected in crayfishless lakes could not be detected in Chozas. The number of orders absent from the macroinvertebrate inventory is high (Table 3a) with the complete depletion of all Heteroptera and Trichoptera species as well as the entire Gastropoda and Hyrudinea Classes.

It needs to be pointed out that the total macroinvertebrate biomass in Chozas is barely 2% of the mean biomass values in other lakes and that more than 70% of this is exclusively due to Odonata.

Amphibians

The herpetological studies carried out in 1980 by Alvarez and Salvador (1984) record breeding of four species of Anura in Chozas lake: *Hyla arborea*, *Pelobates cultripes*, *Bufo calamita* and *Rana perezi*, and two species of urodeles, *Triturus marmoratus* and *Pleurodeles waltl*, the latter being endemic in the Iberian Peninsula and Morocco.

In the aquatic fauna samplings carried out after the arrival of red crayfish in the lake, no evidence of breeding by any of the Anura species has been confirmed. All the species have become extremely rare (isolated adult specimens of *R. perezi* and *H. arborea* have been detected) or have not been detected (*B. calamita* and *P. cultripes*) nor has it been possible to find evidence of laying, larval stages or adults of *T. marmoratus* there; however, the sharp-ribbed salamander (*P. waltl*) population maintains numerous adults and laying has been observed in the remaining vegetation area.

The data on sharp-ribbed salamander abundance in Chozas, measured as captures per unit of effort (CPUE), show values similar to those of other lagoons in the area not colonised by any crayfish species. CPUE of one individual in Chozas are comparable to those observed in Redos (two individuals) or Sentiz (one individual), both eutrophic lakes dominated by vegetation and without the presence of crayfish, although values of CPUE higher than for 90 individuals have also been recorded in the Villaverde lake, which is similar to the two previously mentioned ones. On the other hand, the CPUE values in the Villadangos lake, which has crayfish and no submerged vegetation, were similar to those of Chozas (one individual).

Water birds

The bibliographical study of the papers on water birds carried out in Chozas prior to 1997 confirms the presence of 50 species in the lake, besides at least another five considered accidental visitors (Table 4). Eleven species used the area for nesting prior to 1997; *Aythia nyroca*, catalogued as SPEC 1 (Tucker and Heath 1994) and considered at risk on a national level (Blanco and González 1992) was censused too. Due to the significance of the lapwing (*Vanellus vanellus*) breeding community, the lake could be considered internationally important (Purroy 1990), according to criteria proposed by Scott (1980). Due to this richness, the lake was included in the Regional Catalogue of Wet Areas of Interest, and a report was written requesting the constitution of an

Table 4. (a) Number of species of birds before and after crayfish introduction in Chozas lake. (b) Anatidae species detected in Chozas before 1997 and their changes in status since crayfish introduction (* decreased populations).

	Before crayfish intro.	After crayfish intro.	% Disappeared
Panel a			
Herbivories and diving ducks	12	7	42
Shorebirds	17	12	25
Ciconiform (stork and allies)	4	4	0
Breeding	11	4	64
Wintering	6	4	34

Panel b	Status in Chozas	After 1997
Anas penelope	Migrant	Not detected
Anas crecca	Breeding	Not detected
Anas platyrhynchos	Breeding	*
Anas acuta	Migrant	Not detected
Anas querquedula	Breeding	Not detected
Anas clypeata	Migrant	*
Aythia ferina	Migrant	Not detected
Aythia nyroca	Migrant	Not detected

ornithological reserve in the lagoon and the adjacent wetland (Purroy 1990).

Although the disappearance of 52% of the species and fall in the population of most bird groups still present have been confirmed, it has to be pointed out that not all the groups have been equally affected (Table 4a). In censuses after 1998 among the shorebirds, environmental change has resulted in the absence of around 32% of the species which frequently appeared in Chozas according to Marcos et al. (1995). Although the numbers have fallen, 50% of them have remained in the lake after the arrival of *P. clarkii*. The chance visitor species or those with large yearly variations have not been included in the calculation of these percentages.

Alegre et al. (1991) report the use of Chozas by 11 species of water birds for nesting. Since the introduction of the red swamp crayfish the absence of breeding in seven species has been noticed; on the other hand, the number of nesting pairs of species dependent on marsh vegetation with floating nests (four species) has decreased, highlighting a 65% fall in the number of coots (*Fulica atra*). The lapwing breeding population is maintained although with a tendency to decrease.

The functional group most affected by the introduction of crayfish has been waterfowl which use aquatic food resources (surface and diving ducks, coot). Comparing the census of over-wintering species before introduction (Alegre et al. 1991) with the present situation confirms the loss of 50% of species, all of them Anatidae; this Familia seems to be the most affected as the censuses after the introduction of crayfish show that 75% of the eight species recorded by Purroy (1990) are absent (Table 4b).

A series of birds whose presence has increased in this system since the arrival of *Procambarus clarkii* requires a separate mention. Among the Ardeidae, the increase in common herons (*Ardea cinerea*) and the regular presence of cattle egret (*Bulbucus ibis*) is well known, although the total abundance of *Egretta garzetta* seems to have decreased slightly. Likewise, the common stork (*Ciconia ciconia*) populations feeding in Chozas have increased. Adding to this, the presence of a new predator diving species during the winter and beginning of spring, the great cormorant (*Phalacrocorax carbo*) has been recorded.

Discussion

The profound changes occurring in Chozas after the disappearance of aquatic vegetation are in keeping with the theory of alternative stable states in shallow lakes (Scheffer 1990). However, in this case, the mechanism of change from a clear to turbid state has been the mechanical destruction of submerged vegetation as a result of the benthic activity of an invading alien crayfish species, the red swamp crayfish *Procambarus clarkii*. Rodríguez et al. (2003) showed that exclusion of *P. clarkii* from certain areas of the lake using mesocosms allowed plant coverage to reach up to 95%. These experiments showed that the seed bank has the potential to recover vegetation spontaneously in the absence of red crayfish but the predator activity of the crayfish prevents it. Experiments on the inclusion of crayfish densities close to the estimated one in Chozas (see Rodríguez et al. 1993) in areas with 95% vegetation cover demonstrated the destructive capacity of this species, as 60% of the plant biomass was destroyed in 2 weeks.

This and other studies (Anastáçio and Marques 1997; Vila-Escalé et al. 2002) show that even relatively low crayfish densities (lower than 1 ind/m^2) can completely remove submerged vegetation from shallow lakes and streams in the Iberian Peninsula. It is also relevant that other studies within this area point to this species' preference for fresh plant food (Gutiérrez-Yurrita et al. 1998) whilst studies carried out in other places indicate that detritus is the main ingredient of *P. clarkii*'s diet (Bernardo and Ilhéu 1994). There is another reason for specifying the geographical location when determining the disturbing potential of invading decapods and that is the fact that there are no autochthonous predator species in the Spanish lakes which can affect the crayfish populations in any way. Another interesting aspect is that autochthonous white clawed crayfish (*A. pallipes*) do not seem to have the herbivore intensity of the American red crayfish. Studies in lakes with abundant macrophyte vegetation (*Chara* sp.) have been shown to host high densities of autochthonous crayfish (C.F. Rodriguez et al., in preparation).

Various studies (Stein 1977; Rabeni 1992; Dorn and Mittelbach 1999) report that in the areas of origin of most crayfish species (generally America) the populations are subject to predation by different species of fish. Given that some of these fish species have also been introduced into various places in the world, it would be interesting to confirm whether the interaction of these exotic species (see Elvira et al. 1996) into their new systems allows these effects to be reduced.

Decreased macroinvertebrate populations as a result of the increased densities of autochthonous and exotic freshwater crayfish have been repeatedly documented in studies on the trophic role of these species in streams and lakes (Hanson et al. 1990; Lodge et al. 1994; Nyström and Perez 1998; Nyström et al. 2001; Correia 2002, amongst others). These studies have confirmed under both experimental and natural conditions the incidence of crayfish predation on macroinvertebrate fauna, demonstrating that some decapod species (for example, *P. clarkii*) change their diet from plant food and detritus to an animal diet in relation to the availability of macroinvertebrates (Correia 2002). The disappearance rate for macroinvertebrates associated with vegetation recorded in Chozas was very high (72% of Genera), although these groups are not directly related to crayfish predation. Amphibian diet studies comparing this same lake and others with the same environment (Santos et al. 1986) also showed that, prior to the introduction of exotic crayfish, the macroinvertebrate fauna of Chozas was more widely diversified. Therefore, it can be stated that vegetation destruction has had a greater incidence on the macroinvertebrates of Chozas than the direct effects of crayfish predation and even there is a group, Chironomidae, that despite being more vulnerable to crayfish predation due to its benthic habitat, increased their numbers when compared with other crayfishless lakes. This fact seems to be related with the increased naked sediment area as chironomid densities are similar to those found in non-vegetated lakes.

The vulnerability of egg-laying and young amphibians of different species has been documented in various studies (Gherardi et al. in press; Axelsson et al. 1997; Nyström and Abjörsson 1999) and it is highly likely that the predation of *P. clarkii* significantly affects recruitment of all the amphibian species. During samplings, no adult specimens of the Anura species have been detected, except for some specimens of *H. arborea* and *R. perezi* near the lake. About the newts, absence or presence of *T. marmoratus* (neither larvae nor adults) seems to confirm the disappearance of this species from Chozas.

The persistence of *P. waltl* in Chozas could be sustained by its possessing chemical defences in its epithelium. Various amphibians present this type of defence against predators (Petranka et al.

1988; Bridges and Gutzke 1997), mainly fish and other invertebrates, although Crossland (1998) has documented its lower efficacy against crayfish. In Chozas, the crayfish were observed attacking the uropygium of the sharp-ribbed salamanders trapped in the sampling nets, killing but not devouring them; similar behaviour has been documented by Axelsson et al. (1997) who state that *Pacifastacus leniusculus* preys on larvae of *Bufo* sp., endowed with chemical defences in their epithelium, but does not devour them.

These observations demonstrate direct predation seem to have contributed to rarefaction of all the species of anura and *T. marmoratus* in Chozas lake, as well as the probable decrease in breeding potential. However, indirect mechanisms associated with the destruction of vegetation (for example, *H. arborea* males select areas with floating vegetation and avoids areas without vegetation (García et al. 1987)) and with the collapse of the trophic chains based on the macroinvertebrates (the diet of urodeles is based 60% on macroinvertebrates (Santos et al. 1986)) could be the most likely causes of the decrease in amphibian populations.

Unfortunately, since García et al. studies any amphibian populations research have been done in this geographic area so there is any kind of control of our results and, when explaining the dynamics of any amphibian population, the decreasing tendency recorded in amphibian populations all over the world must be considered (Sarkar 1996; Houlanan et al. 2000). Its causes are barely understood, but it has also been detected in most aquatic systems in our latitudes.

The relationship existing between aquatic plant cover and water birds has been repeatedly documented in various studies (Mitchell and Wass 1996; van Donk and Otte 1996; Sondergaard et al. 1998; Blindow et al. 2000). Although many of them have established their work regimes trying to measure the effect of bird herbivorism on the plant communities (see Mitchell and Perrow 1998), there is a good correlation between the presence of plants and abundance of most water birds. In this sense, the studies based on the analysis of long seasonal series of data (Hargeby et al. 1994; van Donk and Otte 1996; Blindow et al. 2000) have allowed a positive correlation of the establishment of clear water states to years of high bird density.

The data recorded in Chozas throughout the last decade are consistent with the results given in the previously mentioned studies, with the special feature of the change from a transparent to a turbid state not being due to a eutrophication process but to the introduction of alien crayfish. Both herbivores (anatids and coots) and other birds have seen a reduction in the available food due to the direct destruction of vegetation and the indirect disappearance of the communities of plant-associated macroinvertebrates. This effect is shown particularly in the population variations among anatids, a well represented group before 1997 and practically non-existent now.

In addition, the effects on the nesting bird communities, apart from the already mentioned reduction in available food sources, can be related synergically to the red crayfish reducing the marshy areas where floating nests are built and only the existence of a marginal lakeside wood has allowed certain nesting to persist in the area (F.J. Purroy, pers. comm.).

It is of interest too, due to its importance for the implementing of environmental policies, that the three species detected before 1997 and classed as being in danger, *Ardeola ralloides, Plegadis falcinellus* and the already mentioned *Athya nyroca,* have not been included in any census after that year. Likewise, the inventory of vulnerable or rare species (Blanco and González 1992) in Chozas before *P. clarkii* was detected to be 10 species (Purroy 1990), of which 70% have not been inventoried again.

The incidence of the shorebird group has been quantitatively lower, perhaps due to the fact that they exploit resources related to the lake shore area, less affected by the presence of crayfish. In another respect, an increase in the Ardeidae and Ciconiidae populations has been observed, which has been recorded in other studies (Barbaressi and Gerardhi 2000; see Gherardi and Holdich 1999). It could be related to the abundance of crayfish populations, as they are a preferential prey of these birds. Similarly, the winter presence of great cormorants (*P. carbo*) in the last few years also benefits from the abundance of exotic crayfish in the aquatic systems (Barbaressi and Gherardi 2000). However, the generalised spread of this species in continental waters of the Iberian Peninsula is being studied and could be due to various causes (De Nie 1995).

Table 5. Summary of main changes in flora and fauna losses in Chozas before and after crayfish introduction (gen: number of genus, sp.: number of species).

	Before crayfish intro.	After crayfish intro.	% Disappeared
Vegetation coverage	95%	<3%	99
Macroinvertebrates (Gen)	31	9	71
Waterfowl (sp.)	50	26	52
Amphibians (sp.)	6	1	83

Reviewing all the taxonomic groups affected by the introduction of red crayfish shows the spread throughout the trophic chain (Table 5), and the direct and indirect effects of crayfish on the lake communities must be differentiated (Figure 1). The crayfish main negative direct effect on ecosystem is the elimination of submerged vegetation and the reduction of macroinvertebrate populations; both effects in fact mean a dramatic depletion of food resources, shelter and laying sites affecting the whole trophic web. Moreover, the change to a turbid state caused by the disappearance of vegetation has a feedback effect reducing light conditions for vegetation development and affecting macrophyte recolonisation even at low crayfish densities. This turbid state seems to be accompanied by an increase in the densities of benthivore fish (Jeppesen and Sammalkorpi 2002), which favours an increase in piscivorous birds such as herons, egrets, storks, grebes and cormorants. In another respect, there seems to be a positive direct effect on its predators (ardeidae and ciconiiformes), which tend to increase in density.

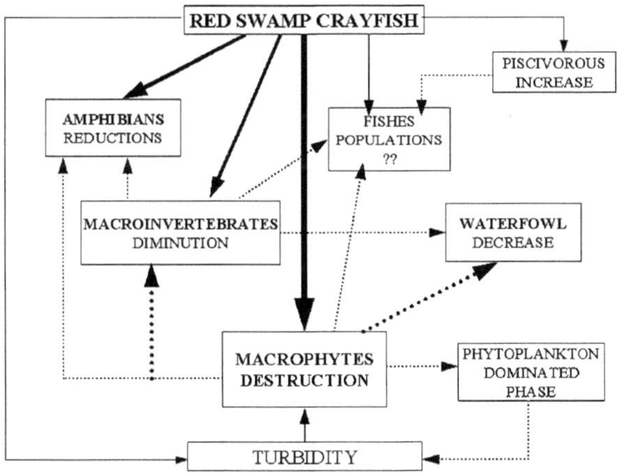

Figure 1. Direct and indirect effects of the exotic red swamp crayfish (*Procambarus clarkii*) on Chozas ecosystem. (continuous arrow: direct effects, dotted arrows: indirect effect).

This study shows that the introduction of the red swamp crayfish *Procambarus clarkii* to the Iberian aquatic systems has dramatic effects on whole ecosystem processes, with direct and indirect repercussions on both flora and fauna communities, mainly as a result of the great predatory effect on submerged vegetation.

Acknowledgements

We wish to thank Benito Fuertes and F.J. Purroy for their comments on water bird fauna. Justo Robles Vicente contributed his own data on the bird census since the introduction of the crayfish. The authors would also like to thank Francisco García and Cristina Trigal for their contributions on macroinvertebrate fauna. This work was funded by the Junta de Castilla-León.

References

APHA-AWWA-WPCF (1989) Estándar Methods for the Examination of Water and Wastewater (Chapter 10), 210 pp

Anastáçio PM and Marques JC (1997) Crayfish, *Procambarus clarkii*, effects on initial stages of rice growing in the lower Mondego River valley (Portugal). Freshwater Crayfish 11: 608–617

Ackerfors H (1999) The positive effects of established crayfish introductions in Europe. In: Gerardhi F and Holdich DH (eds) Crayfish in Europe as Alien Species. Crustacean Issues 11, pp 49–61. A.A. Balkema, Rotterdam, The Netherlands

Angeler DG, Sánchez-Carrillo S, García G and Alvarez-Cobelas M (2001) The influence of *Procambarus clarkii* (Cambaridae, Decapoda) on water quality and sediment characteristics in Spanish floodplain wetland. Hydrobiologia 464: 89–98

Axelsson E, Nyström P, Sidenmark J and Brönmark C (1997) Crayfish predation on amphibian eggs and larvae. Amphibia-Reptilia 18: 217–228

Alegre J, Hernández A and Velasco T (1991) Las aves acuáticas en la provincia de León. Alegre, Hernández y Velasco Eds, León, 78 pp

Álvarez J and Salvador A (1984) Cría de anuros en la laguna de Chozas de Arriba (León) en 1980. Mediterránea 7: 27–48

Barbaresi S and Gherardi F (2000) The invasion of the alien crayfish *Procambarus clarkii* in Europe, with particular reference to Italy. Biological Invasions 2: 259–264

Blanco JC and González JC (1992) Libro rojo de los vertebrados españoles. ICONA, Madrid, 714 pp

Bernardo JM and Ilhéu M (1994) Red swamp crayfish (*Procambarus clarkii)*: Contribution to material cycling. Verh. Int. Ver. Limnol. 25: 2447–2449

Blindow I, Andersson G, Hargeby A and Johansson S (1993) Long term pattern of alternative stable states in two shallow eutrophic lakes. Freshwater Biology 30: 159–167

Bridges CM And Gutzke WH (1997). Effects of environmental history, sibship and age on predator-avoidance responses of tadpoles. Canadian Journal of Zoology 75: 87–93

Carpenter SR and Lodge DM (1986) Effects of submerged macrophytes on ecosystem processes. Aquatic Botany 26: 341–379

Correia AM (2002) Niche breadth and trophic diversity: feeding behaviour of the red swamp crayfish (procambarus clarkii) towards environmental availability of aquatic macroinvertebrates in a rice field (Portugal). Acta Oecologica 23: 421–429

Cronin G (1998) Influence of macrophyte structure, nutritive value and chemistry on the feeding choices of a generalist crayfish. In: Jeppesen E, Sondergaard M, Sondergaard M and Christoffersen K (eds) The Structuring Role of Submerged Macrophytes in Lakes, pp 149–174. Springer-Verlag, New York

Crossland MR (1998) Ontogenic variation on toxicity of tadpoles of the introduced toad Bufo marinus to native Australian aquatic invertebrate predators. Herpetologica 54: 364–369

De Nie H (1995) Changes in the inland fish populations in Europe in relation to the increase of the cormorant Phalacrocórax carbo sinensis. Ardea 83: 115–122

Dorn NJ and Mittelbach GG (1999) More than predator and prey: a review of interactions between fish and crayfish. Vie et Milieu – Life and Environment 49(4): 229–237

Dieguez-Uribeondo J and Soderhall K (1993) Procambarus clarkii Girard as a vector for the crayfish plague fungus, Aphanomyces astaci Schikora. Aquaculture and Fisheries management 24(6): 761–765

Elvira B, Nicola GG and Almodóvar A (1996) Pike and red swamp crayfish: A new case on predator–prey relationship between aliens in central Spain. Journal of Fish Biology 48(3): 437–446

Feminella JW and Resh VH (1989) Submerged macrophytes and grazing crayfish: an experimental study of herbivory in a California freshwater marsh. Holartic Ecology 12(1): 1–8

Fernández-Aláez M (1984) Distribution of macrophytes and their ecological factors in lentic systems from León Province (Spain). Thesis, Departamento de Ecología, Genética y Microbiología, Universidad de León, Spain [in Spanish]

Fernández-Aláez M, Fernández-Aláez C, Rodriguez S and Bécares E (1999) Evaluation of the state of conservation of shallow lakes in the province of León (Northwest Spain) using botanical criteria. Limnetica 17: 107–117

Fernández-Aláez M, Fernández-Aláez C and Rodríguez S (2002) Seasonal changes in biomass of charophytes in shallow lakes in the northwest of Spain. Aquatic Botany 72: 335–348

García C, Salvador A and Santos FJ (1987) Ecología reproductiva en una charca temporal de León (Anura: Hylidae). Rev. Esp. Herp. 2: 33–47

García-Criado F, Fernández-Aláez M, Fernández-Aláez C and Bécares E (2004) Plant-associated macroinvertebrates and ecological status of shallow lakes in north-west Spain. Arch. Hydrobiol. (in press)

Gherardi F and Holdich DM (1999) (eds) Crayfish in Europe as alien species. Crustacean Issues 11. A.A. Balkema, Rotterdam, The Netherlands

Gherardi F, Renai B and Cozti C (in press) Crayfish predation on tadpoles: a comparison between native (Austrapotamobius pallipes) and an alien species (Procambarus clarkii). Bulletin Français de la Pêche et de la Pisciculture

Guanz RZ and Wiles PR (1997) Ecological impact of introduced crayfish on benthic fishes in British lowland rivers. Conservation Ecology 11(3): 641–647

Gutiérrez-Yurrita PJ and Montes C (1999) Bioenergetics and phenology of reproduction of the introduced red swamp crayfish, *Procambarus clarkii*, in Doñana National Park, Spain, and implications for species management. Freshwater Biology 42: 561–574

Gutiérrez-Yurrita PJ, Sancho G, Bravo MA, Baltanás A and Montes C (1998) Diet of the red swamp crayfish Procambarus clarkii in natural ecosystems of the Doñana National Park temporary freshwater marsh (Spain). Journal of Crustacean Biology 18(1): 120–127

Hanson JM, Chambers PA and Prepas EE (1990) Selective foraging by the crayfish Orconectes virilis and its impact on macroinvertebrates. Freshwater Biology 24: 69–80

Hargeby A, Andersson G, Blindow I and Johansson S (1994) Trophic web structure in a shallow eutrophic lake during a dominance shift from phytoplankton to submerged macrophytes. Hydrobiologia 279/280: 83–90

Houlahan JE, Findlay CS, Schmidt BR and Meyer AH (2000) Quantitative evidence for global amphibian population declines. Nature 404: 752–755

Hill AM and Lodge DM (1999) Replacement of resident crayfishes by an exotic crayfish: the roles of competition and predation. Ecological Applications 9(2): 678–690

Jeppesen E and Sammalkorpi I (2002) Lakes. In: Perrow M and Davy T (eds) Handbook of Ecological Restoration; Vol 2: Restoration Practice, pp 297–324. Cambridge University Press, Cambridge

Jeppesen E, Jensen JP, Kristensen P, Sondergaard M, Mortensen E, Sortkjaer O and Olrik K (1990) Fish manipulation as a lake restoration tool in shallow, eutrophic, temperate lakes 2: treshold levels, long-term stability and conclusions. Hydrobiologia 200/201: 219–227

Jeppesen E, Sondergaard M and Christofferson K (eds) (1997) The Structuring Role of Submerged Macrophytes. Springer-Verlag, New York, 423 pp

Krebs CJ (1991) Ecological Methodology. Addison Wesley, Madrid

Lodge DM, Cronin G, van Donk E and Froelich AJ (1998) Impact of herbivory on plant standing crop: comparisons among biomes, between vascular and non-vascular plants, and among freshwater herbivore taxa. In: Jeppesen E, Sondergaard M, Sondergaard M. and Christoffersen K (eds) The Structuring Role of Submerged Macrophytes in Lakes, pp 149–174. Springer-Verlag, New York

Mitchell SF and Perrow MR (1998) Interactions between grazing birds and macrophytes. In: Jeppesen E,

Sondergaard M, Sondergaard M and Cristoffessen K (eds) The Structuring Role of Submerged Macrophytes in Lakes, pp 175–196. Springer-Verlag, New York

Mitchell SF and Wass RT (1996) Grazing by black swans (*Cygnus atratus Latham*), physical factors, and the growth and loss of aquatic vegetation in a shallow lake. Aquatic Botany 55: 205–215

Moss B (1990) Engineering and biological approaches to the restoration from eutrophication of shallow lakes in which aquatic plant communities are important components. Hydrobiologia 200/201: 367–377

Nyström P (1999) Ecological impact of introduced and native crayfish on freshwater communities: European perspectives. In: Gherardi F and Holdich DM (eds) Crayfish in Europe as Alien Species, pp 63–85. A.A. Balkema. Rotterdam, The Netherlands

Nyström P and Abjörnsson K (1999) Effects of chemical cues on the interactions between tadpoles and crayfish. Oikos 87

Nyström P and Pérez JR (1998) Crayfish predation on the common pond snail (*Lygnea stagnalis*): the effect of habitat complexity and snail size on foraging efficiency. Oikos 88(1): 181–190

Nyström P, Svensson O, Lardner B, Brönmark C and Granéli W (2001) The influence of multiple introduced predators on a littoral pond community. Ecology 82: 1023–1039

OECD (Organization for Economic Cooperation and Development) (1982) Eutrophication of waters. Monitoring, assessment and control. Final report OECD Cooperative Programme on Monitoring of Inland Waters (Eutrophication Control); Environment Directorate. OECD, Paris, 154 pp

Olsen TM, Lodge DM, Capelli GM and Houlihan RJ (1991) Mechanism of impact of an introduced crayfish (*Orconectes rusticus*) on littoral congeners, snails, and macrophytes. Canadian Journal of Fisheries and Aquatic Sciences 48(10): 1853–1861

Parkyn SM, Collier KJ and Hicks BJ (2001) New Zealand stream crayfish: functional omnivores but trophic predators? Freshwater Biology 46: 641–652

Petranka JW, Kats LB and Sih A (1987) Predator–prey interactions among fish and larval amphibians: use of chemical cues to detect predatory fish. Animal Behaviour 35: 420–425

Purroy FJ –coord– (1990) Propuesta para establecer una reserva ornitológica en la laguna de Chozas de Arriba. Unpublished report

Rabeni CF (1992) Trophic linkage between stream centrarchids and their crayfish prey. Canadian Journal of Fish and Aquatic Science 49(8): 1714–1721

Rodríguez CF, Bécares E and Fernández-Aláez M (2003) Shift from clear to turbid phase in Lake Chozas (NW Spain) due to the introduction of American red swamp crayfish (*Procambarus clarkii*). Hydrobiologia 506–509: 421–426

Sarkar S (1996) Ecological theory and anuran declines. BioScience 46: 199–207

Santos FJ, Salvador A and García C (1986) Dieta de larvas de *Pleurodeles waltl* y *Triturus marmoratus* (Amphibia: Salamandridae) en simpatria en dos charcas temporales de León. Revista Española de Herpetología 1: 295–313

Scheffer M (1990) Multiplicity of stable states in fresh water ecosystems. Hydrobiologia 200–201: 475–486

Scheffer M and Jeppesen E (1998) Alternative Stable States. In: Jeppesen E, Sondergaard M, Sondergaard M and Cristoffessen K (eds) The Structuring Role of Submerged Macrophytes in Lakes, pp 397–406. Springer-Verlag, New York

Scott DA (1980) A Preliminary Inventory of Wetlands of International Importance for Waterfowl in Western Europe and Northwest Africa. IWRB Publ., Slimbridge, UK

SEO/Birdlife (1997) Atlas de las aves de España (1975–1995), pp 164–165. Lynx Ed, Barcelona, Spain

Soderback B (1994) Interactions among juveniles of 2 freshwater crayfish species and a predatory fish. Oecologia 100(3): 229–235

Sondergaard M, Lauridsen TL, Jeppesen E and Bruun L (1998) Macrophyte-Waterfowl interactions: tracking a variable resource and the impact of herbivory on plant growth. In: Jeppesen E, Sondergaard M, Sondergaard M and Cristoffessen K (eds) The Structuring Role of Submerged Macrophytes in Lakes, pp 298–306. Springer-Verlag, New York

Stein RA (1977) Selective predation, optimal foraging and predator–prey interaction between fish and crayfish. Ecology 58(6): 1237–1253

Tucker GM and Heath MF (eds) (1994) Birds in Europe: their conservation status. Birdlife International. Cambridge

Vandergaag MA (1992) Combined effects of chemicals: an essential element in risk extrapolation for aquatic ecosystems. Water Science and Technology 25 (11): 441–447

Van Donk E and Gulati RD (1995) Transition of a lake to turbid state six years after biomanipulation: mechanisms and pathways. Water Science Technology 32: 197–206

Van Donk E and Otte A (1996) Effects of grazing by fish and waterfowl on the biomass and species composition of submerged macrophytes. Hydrobiologia 340: 285–290

Velasco T, Hernández A and Alegre J (1988) Las aves limnícolas y su paso migratorio por la provincia de León. Tierras de León 72: 115–133

Vila-Escalé M, Rieradevall M and Prat N (2002) Estudio de la población de cangrejo rojo (*Procambarus clarkii*) sobre las comunidades vegetales sumergidas en dos torrentes del macizo de Sant Llorenç del Munt i L'Obac. In: Resúmenes del XI Congreso de la Asociación Española de Limnología y III Congreso Ibérico de Limnología, 17–21 June 2002. Cedex Publ., Madrid

Biological Invasions (2005) 7: 87–97

Worldwide invasion of vector mosquitoes: present European distribution and challenges for Spain

Roger Eritja[1,7,*], Raúl Escosa[2,7], Javier Lucientes[3,7], Eduard Marquès[4,7], Ricardo Molina[5,7], David Roiz [5,7] & Santiago Ruiz[6,7]

[1]*Servei de Control de Mosquits, Consell Comarcal del Baix Llobregat, Parc Torreblanca, 08980 Sant Feliu de Llobregat, Spain;* [2]*CODE, Consell Comarcal del Montsià, Avenida I. Soriano Montagut 86, 43870 Amposta, Spain;* [3]*Departamento de Parasitología, Facultad de Veterinaria, Universidad de Zaragoza, Miguel Servet 177, 50013 Zaragoza, Spain;* [4]*Servei de Control de Mosquits de la Badia de Roses i Baix Ter, Plaça del Bruel 1, 17486 Castello d'Empuries, Spain;* [5]*Instituto de Salud Carlos III, Centro Nacional de Microbiología, Unidad de Parasitología, Carretera de Majadahonda Km2, 28220 Madrid, Spain;* [6]*Servicio de Control de Mosquitos, Diputación de Huelva, Martín Alonso Pinzón 9, 21003 Huelva, Spain;* [7]*EVITAR multidisciplinary network for the study of viruses transmitted by arthropods and rodents;* **Author for correspondence (e-mail: reritja@elbaixllobregat.net; fax: +34-936-300142)*

Received 4 June 2003; accepted in revised form 30 March 2004

Key words: *aegypti, albopictus, atropalpus*, dengue, Europe, invasive, *japonicus*, mosquito, Spain, vector, yellow fever

Abstract

An Asiatic mosquito species, *Aedes albopictus*, began to spread worldwide in the 1970s thanks to marine transport of tires and other goods, leading to colonization of many areas of the world. This species is a vector of major human diseases such as Dengue, Yellow Fever and the West Nile virus. In Europe, it was established in Albania and Italy and has been detected in other countries such as France; no records exist for Spain as yet. Colonization by *Aedes albopictus* is a major public health concern considering that the West Nile virus and several other viruses are known to circulate sporadically in the Mediterranean. Additionally, the parent species *Aedes aegypti* was the vector causing severe outbreaks of Dengue and Yellow Fever two centuries ago. Although *Ae. aegypti* was also introduced, it was eradicated from Spain. Both mosquitoes shared habitat types, diseases transmitted and many bionomic data. This article contains a review of the present *Ae. albopictus* distribution range worldwide and discusses the likelihood of an establishment in Spain in view of climatological and geographical data.

Introduction

Globalizing the economy leads to an increase of the worldwide transport of goods, which raises the chances of accidental transport of foreign species. This has been the case of many agricultural pests unknowingly embarked within plant shipments, leading occasionally to establishment in destination countries and challenging local economies as well as natural systems.

Other groups of species play an important role in public health. Mosquitoes are vectors of many relevant human diseases, from Malaria to filariasis as well as viral pathogens such as Dengue, Yellow Fever and the West Nile virus. Therefore, foreign mosquito species entering new countries may not only produce ecological stress but they are also considered a potential threat to public health. The most notorious case in the past was the ship-mediated introduction in the Mediterranean area of *Aedes aegypti*, causing Yellow Fever and Dengue outbreaks during the 18th and 19th centuries.

Most of the present concern on the invasion of temperate areas by tropical vectors is focused on accidental transport of infected insects from tropical countries on aircraft (Isaäcson 1989). However, commercial worldwide transport of used tires, for example, is an efficient carrier for some mosquito species, so there is a need again to monitor marine transport as a potential threat to human health.

The present article deals with the unprecedented, rapid worldwide spread of the vector mosquito *Aedes (Stegomyia) albopictus* (Skuse 1894) (Diptera: Culicidae), from its original areas in Asia to the rest of the world through colonization of shipments of used tires. There are a number of excellent reviews on *Ae. albopictus,* commonly referred to as the 'Asian Tiger Mosquito' (see e.g. Hawley 1988; Mitchell 1995), so we will mostly focus on reviewing the present European situation and its implications for Spain.

Bionomics

Aedes albopictus is a treehole mosquito, and so its breeding places in nature are small, restricted, shaded bodies of water surrounded by vegetation. However, its ecological flexibility allows it to colonize many types of man-made sites such as cemetery flower pots, bird baths, soda cans, abandonded recipients and especially used tires. As these are often stored outdoors, they collect rainfall and retain rain water for a long time. The addition of decaying leaves from the neighboring trees produces chemical conditions similar to tree holes, thus providing an excellent substitute breeding place. It has been pointed however that *Ae. albopictus* can also establish and survive throughout non-urbanized areas lacking any artificial containers, raising additional public health concerns if mosquitoes are likely to come into contact with enzootic arbovirus cycles (Moore 1999). The adult flight range is quite short, as expected for a scrub-habitat mosquito. Therefore, most medium and long range colonization is the result of passive transportation.

Aedes albopictus is an aggressive, outdoor daytime biter that attacks humans, livestock, amphibians, reptiles and birds. The females lay desiccation-resistant eggs above the surface of the water in treeholes or tires. The eggs from strains colonizing temperate regions resist lower temperatures than those from tropical areas (Hanson and Craig 1995). Additionally, in these strains, the combination of short photoperiods and low temperatures can induce the females to lay diapausing eggs which can hibernate (Hanson and Craig 1995). Overwintering is necessary north of the +10 °C January isotherm (Mitchell 1995; Knudsen et al. 1996). The combination of these adaptations accounts for the success in colonizing temperate regions.

Recent spreading and present distribution

A fraction of the present Asiatic distribution range of *Ae. albopictus* is the result of invasions prior to the 20th century, as in Hawaii before 1902 (Sprenger and Wuithiranyagool 1986).

The first modern establishment outside this original range occurred in 1979 in Albania (Adhami and Reiter 1998) although not much concern was raised due to the political isolation of the country. The species is believed to have already been there for some years when discovered, and was probably imported in tire shipments from China (Adhami and Reiter 1998).

Aedes albopictus was next detected in the United States in 1985. Although scattered individuals had already been sporadically collected in the country (Hawley 1988; Reiter 1998), the cluster detected in Harris County, Texas, was the first established population (Sprenger and Wuithiranyagool 1986). Adaptation to cold suggested that the strain probably came from a non-tropical area of Asia, as confirmed by specimen detection in tires coming from Japan (Reiter 1998). In the US, the eastward dispersion of the mosquito was very rapid while the spread to the north and the west was slower, probably due to increasing dryness and cold, respectively (Moore 1999); in 2003, 866 counties from 26 states were infested (CDC, unpublished data).

In 1986 *Ae. albopictus* was detected in Brazil, and Mexico became the next positive country in 1988. Between that year and 1995, the species was detected in most of Central America (Honduras, Costa Rica, Guatemala, El Salvador, Panama), part of the Caribbean islands after 1993 (firstly Dominican Republic, then Cayman Islands and Cuba). More recently, *Ae. albopictus* has also been reported from Guatemala and Bolivia (1995), Colombia (1997), Argentina (1998) and Nicaragua (2003).

In the Pacific area, *Ae. albopictus* was detected in Salomon, Australia (1988), Fidji (1988), New Zealand (1994), and La Réunion (1994). Some African countries such as South Africa (1990) have detected the species, with establishment in Nigeria in 1991. It was recently found to be well established in southern Cameroon (Fontenille and Toto 2001). No other African country has reported *Ae. albopictus,* but the scarcity of surveys might mask a broader presence in the continent.

European concern rose when the species was detected in Italy, firstly in September 1990 as a few adults of unknown origin in Genoa (Sabatini et al. 1990). An established population was found 1 year later near Padua (Dalla Pozza and Majori 1992). Research disclosed that the infestation originated in a tire depot that received egg-infested shipments of aircraft tires from Atlanta, US. Further genetic analysis showed affinities between Italian, US and Japanese *Ae. albopictus* (Urbanelli et al. 2000). The Tiger Mosquito rapidly spread across the northern and central regions of Italy and Sardinia by means of domestic tire trading, and reached Rome in 1997; although some local eradications have been achieved, the species is now present in nine regions and 190 municipalities (Romi 2001).

Ae. albopictus was first found in two tire dumps in France in 1999 during a specific survey (Schaffner and Karch 2000; Schaffner et al. 2001). There was evidence that the species was established from at least the previous year. Chemical control actions undertaken in 2001 by health authorities apparently eradicated the mosquito from these points (Schaffner 2002). However, the presence of *Ae. albopictus* was detected the same year in a new continental location and in Corsica as well by 2002 (F. Schaffner, pers. comm.).

Investigating the French findings and tracing back the route of infested shipments led to the discovery of *Ae. albopictus* in the year 2000 in one location in Belgium, which became the fourth European positive country (Schaffner 2002). Recently, the species has been formally reported from Montenegro (D. Petric, pers. comm.) and from discarded tires in the vicinity of an airport in Israel (Pener et al. 2003). There are references on the possible presence of *Ae. albopictus* in Hungary (Schaffner 2002), but no direct reports.

It is worth noting that many detection reports of *Ae. albopictus* have not been followed by establishment. Quarantine and inspection measures in Australia allowed detection of 17 larval introductions between 1997 and 2001 and five more interceptions in seaports since 2001. As immediate control measures have been applied, *Ae. albopictus* has not yet become established in the continent (R. Russell, pers. comm.). In the Mediterranean, only the introductions in Albania, Italy and probably Israel led to establishment. The other cases are very local, too recent or have been subject to eradication actions yet to be evaluated.

New transport types and other mosquitoes

During the summer of 2001, containerized shipments from China of the plant known as Lucky Bamboo (*Dracaena* spp.) were found to contain *Aedes albopictus* on inspection by quarantine officers on arrival at Los Angeles, USA (Linthicum 2001). Several live adult mosquitoes escaped while opening as *Ae. albopictus* larvae had been transported within *Dracaena* plants shipped in standing water. Destination wholesale nurseries in California were also found to be infested (Madon et al. 2002).

The trade in Lucky Bamboo is increasing because it has cultural relevance within the Asiatic communities in the US and elsewhere, and it has also gained worldwide attention as a popular gift. Large nurseries are located in the Guangdong province of China, where the climate is suitable for *Ae. albopictus* (Madon et al. 2002).

Although the problem appeared recently, the importation of Lucky Bamboo plants is not recent. However, until ca. 1999, the plants were dry packaged and airfreighted; the increase in demand and cost cutting led to the use of container ships. The plants are usually transported in standing water, thus providing the conditions for the breeding of the *Ae. albopictus* larvae. Therefore, the US authorities dictated an embargo on this type of shipment favoring dry airfreight. However, this overlooked the possibility of mosquito eggs being transported on the plant stems.

The spread of *Ae. albopictus* might well be only the first step in mosquitofauna globalization. Similar mechanisms that allowed invasion by *Ae. albopictus* have also transported other mosquito species. The North American mosquito *Ochlerotatus atropalpus* (Coquillett 1902) was also introduced in Italy through the tire trade from the USA as detected in the Veneto region (Romi et al. 1997). As the infestation was local, rapid control measures greatly reduced the population density in 1997, with no positive reports in 1998 (Romi et al. 1999).

Monitoring *Ae. albopictus* in France led to the discovery of another exotic treehole species: in this case, *Ochlerotatus* (Finlaya) *japonicus japonicus* (Theobald 1901) was found in French territory in 2001 (Schaffner et al. 2003). This mosquito probably came together wth *Ae. albopictus* in tires from the USA, where it is present in several Eastern states following introduction from Japan in 1998 (CDC, unpublished data). Both *Ochlerotatus japonicus* and *Ae. atropalpus* are efficient vectors of the West Nile virus (Turell et al. 2001); and *Oc. japonicus* is also believed to be a vector of Japanese Encephalitis (CDC, unpublished data).

Starting in 1992, several countries in South America (to our knowledge, Venezuela, Chile, Bermuda, Costa Rica, Argentina and Brazil) have dictated embargoes on used tire importations, in an attempt to not only prevent mosquito introduction but also to protect local industries as well as prevent Dengue if *Ae. aegypti* is already present. Although this is an efficient strategy, it also has an economic impact; additionally, in the European Union, it would be a less-efficient measure as due to free internal commerce, the country of origin may remain unknown (Reiter 1998). Several countries have passed regulations for the inspection, certification and quarantine of used tires (Reiter 1998), but these are difficult to enforce thoroughly. Local laws have been passed in Italy, but no tire legislation exists at the national level (Romi et al. 1999).

Competition with other mosquito species

Little, if any, attention has been paid to the impact of the presence of *Ae. albopictus* on autochthonous tree-hole breeding mosquitoes. In Spain, interspecific competition might affect *Aedes (Finlaya) geniculatus* (Olivier 1791), *Ochlerotatus (Ochlerotatus) berlandi* Séguy 1921, *Anopheles (Anopheles) plumbeus* Stephens 1828 and the less frequent *Orthopodomyia (Orthopodomyia) pulcripalpis* (Rondani 1872), among others.

Competition has been studied, however, between imported vectors. Distribution ranges of *Ae. albopictus* and *Ae. aegypti* partially overlap, although they occupy different biotopes. The former inhabits densely vegetated rural environments, whereas *Ae. aegypti* prefers less humid, urban breeding places (Mitchell 1995). In some parts of Asia, a general replacement of *Ae. albopictus* by *Ae.aegypti* has been noted. This may be attributable to the urbanization of rural areas (Hawley 1988). However, *Ae. albopictus* will also readily colonize urban habitats if *Ae. aegypti* is not present. Therefore, it has been suggested that larval competition resolved in favor to *Ae. aegypti* could also play a role in some habitats (Hawley 1988).

Interestingly, the opposite replacement is recorded in certain locations in the USA after the introduction of *Ae. albopictus*, apparently inducing the decline or even disappearance of *Ae. aegypti* (Hobbs et al. 1991). It has been hypothesized that the better adaptations to colder climates by *Ae. albopictus* are a reason for this exclusion. In reviewing the literature, Christophers (1960) concluded that the dominant factor on *Ae. aegypti* distribution was a short summer season rather than low winter temperatures. In the laboratory, tropical *Ae. aegypti* eggs can survive for months at 4 °C (P. Reiter, pers. comm.) However, in regions with a January isotherm close to 0 °C, the air temperature is below freezing for many days in the winter months. Clearly, *Ae. aegypti* eggs can only survive in such conditions if sheltered, but these circumstances are not uncommon: in Memphis, Tennessee, the species was abundant in late spring after a winter in which temperatures had dropped to −18 °C (P. Reiter, pers. comm.). Thus, factors other than temperature have induced these changes of range.

Due to this replacement of species, the arrival of *Ae. albopictus* has sometimes been hailed as good public health news because *Ae. aegypti* is considered to be more efficient as a Dengue vector. Unfortunately, *Ae. albopictus* was more receptive to artificial laboratory infection with the West Nile virus than *Ae. aegypti* (Turell et al. 2001).

Aedes aegypti-related diseases in Spain

The Mediterranean Yellow Fever and Dengue outbreaks during the 19th century were transmitted by the parent species *Ae. aegypti,* present as a result of previous invasions. Both species share much of habitat types, bionomics and vector diseases; thus, information from the past distribution of *Ae. aegypti* is worth considering here.

Earlier outbreaks of Yellow Fever in Spain occurred from 1701 onwards. The disease especially affected the southernmost region of the country, where it remained endemic for more than a century (Pittaluga 1928). Both the vector and the disease were imported by sailboats, so outbreaks originated in coastal cities reaching further inland locations, sometimes as far as Madrid (Pittaluga 1928). A single concatenation of Yellow Fever outbreaks in 1800–1803 took >60,000 lives in Cádiz, Sevilla and Jerez de la Frontera (Nájera 1943; Angolotti 1980). According to Pittaluga (1928), another episode in Barcelona (1822–1824) affected 80,000 inhabitants, 20,000 of whom died. There are total estimates of more than 300,000 casualties from Yellow Fever during the first half of the 19th century; a detailed epidemiologic review can be found in Rico-Avelló (1953). The last Yellow Fever episodes in Spain occurred between 1870 and 1880 (Nájera 1943).

The name 'Dengue' is derived from the Spanish word 'derrengue', which applies to a condition of extreme exhaustion (Angolotti 1980); the word is still used in parts of southwestern Spain as an adjective for lazy people (T. Romero, pers. comm.). Dengue epidemics were not as well documented as Yellow Fever, but can be traced to southern and eastern Spain; the first probable outbreak is recorded in Cádiz in 1778. The mortality was so low that the disease was popularly called 'La Piadosa' ('the compassionate') (Angolotti 1980). Although physicians were aware of the different nature of the two diseases, Dengue was less noticed than Yellow Fever because it caused much lower mortality. An outbreak from 1927 reported by Pittaluga (1928) killed less than 5% of infected people, simultaneously to the huge outbreak in Greece that caused one million cases in 2 years, of which more than 1000 died (Adhami and Reiter 1998).

The last documented sample of *Ae. aegypti* was collected in downtown Barcelona in 1939 (Margalef 1943), and the species was described as 'very common'. In his review on the Aedines of Spain, Clavero (1946) also quoted *Ae. aegypti* as being common, but remarked that the present distribution should be better documented. Rico-Avelló (1953) again considered the species as being 'very common' in Spain (see his map in Figure 1) but failed to list his references. García Calder-Smith (1965) did not find *Ae. aegypti* in the Barcelona province, despite multi-year sampling from 1958 to 1965. More recent reviews by Torres Cañamares (1979) and Encinas Grandes (1982) both stated again that *Ae. aegypti* was present in Spain, again on a bibliographic basis only. Thus, due to a lack of field reports, the position was adopted in the latest checklist on the Spanish mosquitoes (Eritja et al. 2000) to formally consider that *Ae. aegypti* had been eradicated, although the reasons for this remain unknown.

Besides sanitation measures, the repetitive introductions by ships were highlighted as a factor of maintenance of the vector. Sailors were aware that old sailboats were healthier than newly built ships: they leaked so much that

Figure 1. Past distribution of *Aedes aegypti* in Spain (redrawn from Rico-Avelló 1953).

pumping had to be continuous, thus suppressing breeding places onboard (Angolotti 1980). Thus, steamers may have been a major change because they allowed better water management and also shortened the journey across the ocean (Nájera 1943), preventing the development of multiple mosquito generations during the trip, which resulted in the infection of the whole crew.

Additional impacts on this species from the Malaria eradication programs during the first half of the 20th century have been suggested (Samanidou-Voyadjoglou and Darsie 1993; Reiter 2001). Unfortunately, as most of these programs remain undocumented in Spain, the impact of campaigns focused on ricefield Anopheline species on an urban, indoor mosquito cannot be discussed.

Public health risks from *Aedes albopictus*

Public health implications are not trivial as *Ae. albopictus* is only second to *Ae. aegypti* in transmission of Yellow Fever and Dengue. The Tiger Mosquito is believed to act as a secondary Dengue vector in rural environments where human population density is much lower than in cities, so that large outbreaks are not likely to occur; many episodes are not even recorded (Hawley 1988). In some cases, however, the absence of *Ae. aegypti* implicates *Ae. albopictus* in larger epidemics, such as the >100,000 case outbreak in Japan during WWII (Kobayashi et al. 2002). Transovarial transmission of Dengue has been demonstrated in the laboratory for *Ae. albopictus* (Rosen et al. 1983) and has also been verified in field-collected larvae (Moore and Mitchell 1997). European-established strains from Albania (Vazeille-Falcoz et al. 1999) and Genoa, Italy (Knudsen et al. 1996), are receptive to the virus.

The Tiger Mosquito is also an efficient vector for other Flaviviruses such as Japanese Encephalitis and West Nile virus. Several West Nile outbreaks have occurred in the Mediterranean, but the 1996 outbreak in Romania was remarkable, with 453 human cases (Hubálek and Halouzka 1999). Following the introduction of the virus in the USA in 1999, during the single year 2002 a total of 4161 human cases were reported, 277 of which died (CDC, unpublished data). *Aedes albopictus* may be a matter of concern as a bridge vector for the West Nile virus because it inhabits rural areas and has a wide host range including birds, so that it can readily pass enzootic cycles to humans. However, many autochthonous mosquito species (including the ubiquitous *Culex pipiens*) can play an exactly same role. Wild populations of *Ochlerotatus japonicus* have also been found infected by the West Nile virus in the USA (Turell et al. 2001) and experimentally infected with the EEE virus (Sardelis et al. 2002).

Table 1 summarizes the known receptivity of *Ae. albopictus* to pathogenic viruses by experimental laboratory infection, as well as the list of viruses isolated from field-collected individuals. Included are the four quoted Flaviviruses, plus seven Alphaviruses and 10 Bunyaviruses. One additional Flavivirus and two Bunyavirus have neither been tested in the laboratory nor in the field but are known to circulate in the Mediterra-

Table 1. Known virus receptivity in the laboratory for *Aedes albopictus;* viruses isolated from wild mosquito populations, and human pathogenic viruses present in the Mediterranean (compiled from Mitchell 1995; Moore and Mitchell 1997; Gerhardt et al. 2001; Holick et al. 2002).

Virus	Laboratory infection	Field positives	Presence in the Mediterranean
Flavivirus			
Dengue (all 4 serotypes)	*	*	* (past)
West Nile	*	*	*
Yellow Fever	*		* (past)
Japanese Encephalitis	*	*	
Israel Turkey Encephalitis			*
Bunyavirus			
Jamestown Canyon	*	*	
Keystone	*	*	
LaCrosse	*	*	
Oropouche	*		
Potosi	*	*	
Rift Valley fever	*		*
San Angelo	*		
Trivittatus	*		
Cache Valley	*	*	
Tensaw	*	*	
Tahyna			*
Batai			*
Alphavirus			
WEE	*		
EEE	*	*	
VEE	*		
Chikungunya	*		*
Sindbis	*		*
Mayaro	*		
Ross River	*		

nean and to be pathogenic to humans (Mitchell 1995). *Ae. albopictus* is also a vector of filariasis caused by *Dirofilaria immitis* (Nayar and Knight 1999).

Tentative forecasts on spreading

The original distribution area in the North of Asia occasionally reaches the −5 °C isotherm, and may do so in North America (Mitchell 1995). Even assuming the more conservative 0 °C isotherm, this means that the species could become established in northern Europe as far as the southern coast of Sweden and Norway, with most countries at risk (Mitchell 1995). This contrasts with the +10 °C January isotherm that apparently delimits establishment areas of *Ae. aegypti* (Knudsen et al. 1996), as well as diapausing populations of *Ae. albopictus*.

Within this broad range, local establishment would depend on climatic conditions based on temperature, photoperiod, humidity and rainfall. It has been suggested (Mitchell 1995; Knudsen et al. 1996) that areas at risk in Europe would have mean winter temperatures higher than 0 °C, at least 500 mm rainfall and a warm-month mean temperature higher than 20 °C. Rainfall and temperature covary regionally, so higher temperatures are positive for the species as long as the breeding sites do not completely dry out (Alto and Juliano 2001). It is believed that less than 300 mm rainfall per year would make establishment extremely unlikely (Mitchell 1995). This is viewed as reasonable; inspection by the authors of tire depots located in Spanish areas with less than 250 mm, disclosed less than 5% of sampled tires contained water, however, in very small amounts (our own unpublished data from September 2002).

The active season in southwestern US and Japan is from late Spring to early fall (Alto and Juliano 2001). In Rome, larvae are found from March to November, but some females are active until December (Di Luca et al. 2001). This situation is likely to be reproduced in Spain. However, a wide array of scattered climatic areas are affected by mountain ranges as well as maritime and continental influences. For a tentative graphic evaluation of the most suitable regions for an *Ae. albopictus* establishment in Spain, series of climatic information have been plotted in Figures 2–6. All underlying data have been collected from reports of the Instituto Nacional de Meteorología (Font 1983) and the Spanish Ministry of Agriculture (unpublished GIS data). The January 0 °C isotherm in Spain is not relevant to this purpose because it only delimits a few high mountain areas; so that the entire country is primarily at risk under this criterion. The Canary islands have been excluded from the plotting because by geographic configuration the influence of microclimates exceeds general climatic influence at this study scale.

Figure 2 plots the mean annual rainfall rate. Following the literature, only areas receiving more than a yearly minimum of 500 mm have been greyed. However, in our climatic conditions, the rainfall can be heavy but occurs on a seasonal basis, failing to provide continuous breeding places for mosquitoes during the warm season. Thus, we have plotted in Figure 3 the areas with >60 rainy days per year (if >0.1 mm water is recorded). This is intended to correct for stormy-season regions and has been verified by checking against a plot of the >0.5 humidity class region, following UNESCO nomenclature.

Figure 4 deals with mean temperatures. The northern blank area is delimited by the 11 °C all-year isotherm, which Kobayashi et al. (2002) found to delimit *Ae. albopictus* distribution in northern Japan. In Spain, this line very closely matches another suggested climate conditioning factor, the 20 °C warm-month isotherm (not shown). This is an interesting coincidence as these two criteria have been proposed by different authors.

Figure 5 accumulates and presents the previous three climate figures graphically. The dark patches are the regions where all three conditions are simultaneously met; thus, they are also the most suitable areas for *Ae. albopictus*.

Climate-based forecasts are a charming entertainment but are of a very simplistic nature, even using good-quality data. Whereas microclimates cannot be considered at this study scale, they may play a major role. It is however indicated in Figure 5 that many inland territories behind the Spanish eastern coast, including the mid-Ebro valley and large areas of Andalusia, are unsuitable owing to dryness. In the latter, however, the presence of several mountain ranges may provide

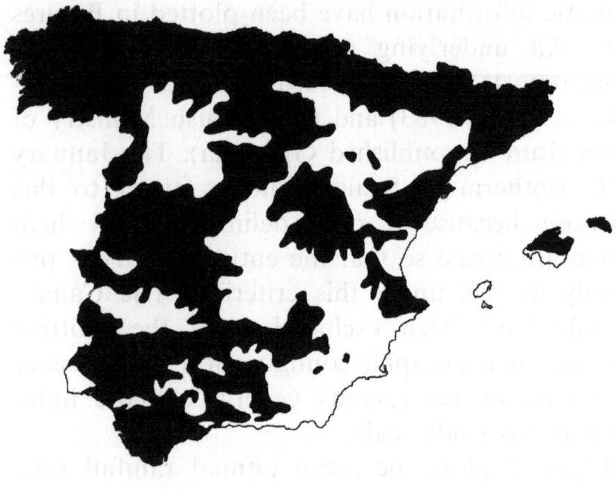

Figure 2. Spanish areas receiving more than 500 mm mean rainfall per year.

Figure 3. Spanish areas having more than 60 rainy days (0.1 mm rainfall minimum each) per year.

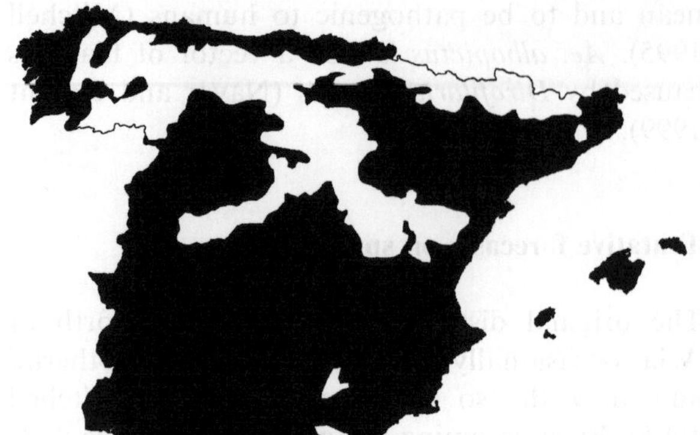

Figure 4. Spanish areas with mean yearly temperatures higher than 11 °C.

Figure 5. Hypothesized suitable areas (darkened) for *Ae. albopictus* establishment, plotted by intersecting dark areas in Figures 2 and 3 and further suppressing from the result the cold (white) area in Figure 4.

suitable conditions within their slopes. Western parts of Extremadura and Leon would be at risk, as well as most of Catalonia, all these areas sharing a relatively dry, warm-summer climate.

On the other hand, it is worth noting that the entire northern Cantabric shore and corresponding inland areas (including also most of Galicia) could allow establishment of *Ae. albopictus*. These areas have more humid and rainy climates. Given that breeding water would not be limiting here, only low local mean temperatures could theoretically prevent establishment of *Ae. albopictus*.

All areas in Figure 5 are re-plotted in Figure 6 against the human population density if there are more than 20 inhabitants per square kilometer. Known tire dumps are also represented in this map by circles. Data on their locations were collected from chambers of Commerce, phone directories, referrals by collaborators and professional societies (unpublished data). This list is only a rough guide because many of the real (and probably more interesting) existing tire depots may not be officially identified as such.

Discussion

Stopping the invasion in the long term is usually considered to be extremely difficult, if not impos-

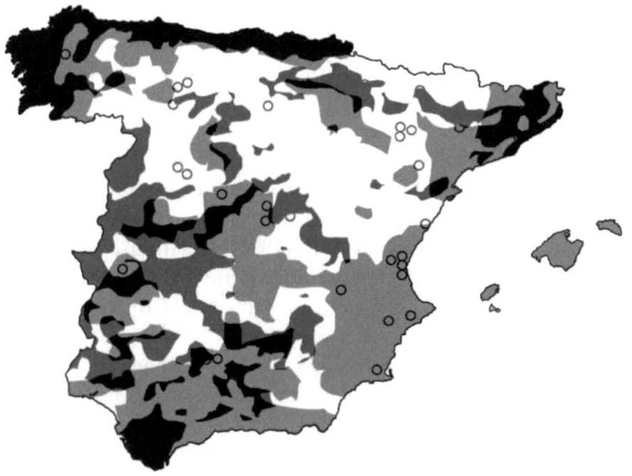

Figure 6. Suitable areas from Figure 5 together with human density population in Spain (grey area; 1991 census, >20 inhabitants per square kilometer). Circles represent the locations of known relevant tire dumps.

sible (Reiter 1998). The spreading of *Aedes albopictus* is quite slow *per se*: it has not yet spread along the Mediterranean coast from Italy to France, in spite of the relatively short distances. All infestations in Italy were tracked to tire depots (Knudsen et al. 1996; Romi et al. 1999), and it was also shown that the early infestations in the US were clustered along the interstate highways (Moore and Mitchell 1997). On the other hand, rapid invasion of some large areas (southern Cameroon, US) strongly suggested multiple simultaneous infestations.

Source reduction strategies such as larval or adult control within tire dumps have proven to be difficult and relatively inefficient due to the shape and abundance of the water surfaces. This was successful in Australia and France (Schaffner 2002) where it has been (and still is) applied on initial invasion stages; source reduction by tire management is more advisable for established situations.

Preliminary data for Spain indicated that used tire importations are a low-volume business, although existing data might underestimate this activity. Export figures collected from origin countries by Reiter (1998) included relevant amounts for Spain as a destination of tire exports from the US, Japan and South Korea between 1989 and 1994. On the other hand, the Lucky Bamboo plant is also being imported to Spain from China. Preliminary test inspections by the authors in February 2002 on *Dracaena* shipments arriving at a wholesale plant nursery disclosed the presence of more than 70 l of standing water in a single container. Plant stems did not have attached eggs, but the remains of one drowned, unidentifiable adult mosquito plus a damaged larva (*Culex* spp.) were filtered out from the water (unpublished data).

Awareness of the risks is absolutely necessary at all official levels even if it is impossible to stop the establishment of *Aedes albopictus* within its suitable geographical range. Such an introduction would be easier to deal with if *Ae. albopictus* could be kept in a rural range, as are the 24 present Aedine species in Spain, none of which occurs significantly within urban environments. In dealing with such aggressive species, the simple biting nuisance can also be a form of public health threat. Preventing the arrival of a new stock and suppressing already present populations would retard their arrival in cities, would limit the replenishment of gene pools and diminish the risk of pathogens introduced within transovarially infected mosquitoes.

The historic relationship between Spain and South and Central America implies many exchanges within these countries. This raises risks derived from the presence of Dengue-infected people that could theoretically initiate transmission in Spain if appropriate mosquitoes were present, as *Aedes aegypti* was two centuries ago. However, a comprehensive healthcare system, housing facilities and many other social factors as well as urban management, would reduce the epidemiological risks such as they may be at present. Monitoring for several viral disease agents would, however, be necessary with a strong emphasis on the West Nile virus as a major local risk. In Spain, no immediate vectorial risks are reasonably expected, and they have not occurred in Italy. However, severe local nuisance could be expected as the experience of Rome clearly demonstrates.

Pittaluga wrote in his documented article on Yellow Fever in Spain and the tropics (1928): 'The problem of the Yellow Fever is a European problem and we must be concerned about the possible danger, taking into account the historical epidemic cycles' [translated by the authors]. These warning words are still valid now that a new potential threat to public health is colonizing

Europe. The Mediterranean received the impact from *Ae. aegypti* and related diseases two centuries ago. At present, all countries are at risk from a parent species that will probably not transmit any significant disease; however, this one came to stay.

In Spain, a scientific network named EVITAR was built up early in 2003 to study and monitor viral arthropod- and rodent-borne diseases. Within this frame, the authors are in charge of the surveillance campaign for managing possible introductions of *Aedes albopictus* and other mosquitoes, as well as other exotic Arthropod species of medical relevance.

References

Adhami J and Reiter P (1998) Introduction and establishment of *Aedes (Stegomyia) albopictus* Skuse (Diptera: Culicidae) in Albania. Journal of the American Mosquito Control Association 14(3): 340–343

Alto BW and Juliano SA (2001) Precipitation and temperature effects on Population of Aedes albopictus (Diptera: Culicidae): implications for range expansion. Journal of Medical Entomology 38(5): 646–656

Angolotti E (1980) La fiebre amarilla. Historia y situación actual. La fiebre amarilla en la Barcelona de 1821. Revista de Sanidad e Higiene Pública 54: 89–102

Christophers SR (1960) *Aedes aegypti* (L.), the Yellow Fever Mosquito. Its life history, bionomics and structure. Cambridge University Press, Cambridge, 739 pp

Clavero G (1946) Aedinos de España. Revista de Sanidad e Higiene Pública XX: 1205–1231

Dalla Pozza G and Majori G (1992) First record of *Aedes albopictus* establishment in Italy. Journal of the American Mosquito Control Association 8(3): 318–320

Di Luca M, Toma L, Severini F, D'ancona F and Romi R (2001) *Aedes albopictus* a Roma: monitoraggio nel triennio 1998–2000. Annali dell'Istituto Superiore di Sanità 37(2): 249–254

Encinas Grandes A (1982) Taxonomía y biología de los mosquitos del área salmantina (Diptera: Culicidae). Consejo Superior de Investigaciones Científicas, Centro de Edafología y Biología Aplicada. Ed. Universidad de Salamanca, Salamanca, Spain, 437 pp

Eritja R, Aranda C, Padrós J, Goula M, Lucientes J, Escosa R, Marquès E and Cáceres F (2000) An annotated checklist and bibliography of the mosquitoes of Spain (Diptera: Culicidae). European Mosquito Bulletin 8: 10–18

Font I (1983) Atlas climático de España. Servicio de Publicaciones del Instituto Nacional de Meteorología. Ministerio de Transportes, Turismo y Comunicaciones, Madrid, 49 pp

Fontenille D and Toto JC (2001) *Aedes (Stegomyia) albopictus* (Skuse), a potential new Dengue vector in Southern Cameroon. Emerging Infectious Diseases 7(6): 1066–1067

García Calder-Smith JR (1965) Estudio de los Culícidos de Barcelona y su provincia. PhD Thesis, Facultad de Farmacia, Universidad de Barcelona, typewritten document, 193 pp + v

Gerhardt R, Gottfried K, Apperson AC, Davis B, Erwin P, Smith A, Panella N, Powell E and Nasci R (2001) First isolation of La Crosse virus from naturally infected *Aedes albopictus*. Emerging Infectious Diseases 7(5): 807–811

Hanson SM and Craig Jr GB (1995) Relationship between cold hardiness and supercooling point in *Aedes albopictus* eggs. Journal of the American Mosquito Control Association 11(1): 35–38

Hawley WA (1988) The biology of *Aedes albopictus*. Journal of the American Mosquito Control Association, 4 (suppl): 39 pp

Hobbs JH, Hughes EA and Eichold BH (2001) Replacement of *Aedes aegypti* by *Aedes albopictus* in Mobile, Alabama. Journal of the American Mosquito Control Association 7(3): 488–489

Holick J, Kyle A, Ferraro W, Delaney RR and Iwaseczko M (2002) Discovery of *Aedes albopictus* infected with West Nile virus in Southeastern Pennsylvania. Journal of the American Mosquito Control Association 18(2): 131

Hubálek Z and Halouzka J (1999) West Nile Fever – a reemerging mosquito-borne viral disease in Europe. Emerging Infectious Diseases 5(5): 643–650

Isaäcson M (1989) Airport Malaria: a review. Bull. WHO 67: 737–743

Knudsen AB, Romi R and Majori G (1996) Occurrence and spread in Italy of *Aedes albopictus*, with implications for its introduction into other parts of Europe. Journal of the American Mosquito Control Association 12(2): 177–183

Kobayashi M, Nihei N and Kurihara T (2002) Analysis of northern distribution of *Aedes albopictus* (Diptera: Culicidae) in Japan by Geographical Information System. Journal of Medical Entomology 39(1): 4–11

Linthicum K (2001) Discovery of *Aedes albopictus* infestations in California. Vector Ecology Newsletter 32(3): 5–6

Madon M, Mulla MS, Shaw MW, Kluh S and Hazelrigg JE (2002) Introduction of *Aedes albopictus* (Skuse) in southern California and potential for its establishment. Journal of Vector Ecology 27(1): 149–154

Margalef R (1943) Sobre la ecología de las larvas de algunos Culícidos (Dípt. Cul.) Graellsia (1): 7–12

Mitchell CJ (1995) Geographic spread of *Aedes albopictus* and potential for involvement in Arbovirus cycles in the Mediterranean basin. Journal of Vector Ecology 20(1): 44–58

Moore CG (1999) *Aedes albopictus* in the United States: current status and prospects for future spread. Journal of the American Mosquito Control Association 15(2): 221–227

Moore CG and Mitchell CJ (1997) *Aedes albopictus* in the United States: ten-year presence and public health implications. Emerging Infectious Diseases 3(3): 329–334

Nájera L (1943) Los aedinos españoles y el peligro de la Fiebre Amarilla. Graellsia I(1): 29–35

Nayar J and Knight J (1999) *Aedes albopictus* (Diptera: Culicidae): an experimental and natural host of *Dirofilaria immitis* (Filarioidea: Onchocercidae) in Florida, USA. Journal of Medical Entomology 36(4): 441–448

Pener H, Wilamowski A, Schnur H, Orshan L, Shalom U and Bear A (2003) Letter to the Editors. European Mosquito Bulletin 14: 32

Pittaluga G (1928) El problema de la Fiebre Amarilla. Medicina de los Países Cálidos, 5–25

Reiter P (1998) *Aedes albopictus* and the world trade in used tires, 1988–1995: the shape of the things to come? Journal of the American Mosquito Control Association 14(1): 83–94

Reiter P (2001) Climate change and mosquito-borne disease. Environmental Health Perspectives 109(1): 141–161

Rico-Avelló C (1953) Fiebre Amarilla en España (Epidemiología histórica). Revista de Sanidad e Higiene Pública 27(1–2): 29–87

Romi R (2001) *Aedes albopictus* in Italia: un problema sanitario sottovaluato. Annali dell'Istituto Superiore di Sanità 37(2): 241–247

Romi R, Sabatinelli G, Savelli LG, Raris M, Zago M and Malatesta R (1997) Identification of a North American mosquito species, *Aedes atropalpus* (Diptera: Culicidae) in Italy. Journal of the American Mosquito Control Association 13(3): 245–246

Romi R, Di Luca M and Majori G (1999) Current status of *Aedes albopictus* and *Aedes atropalpus* in Italy. Journal of the American Mosquito Control Association 15(3): 425–427

Rosen L, Shroyer DA, Tesh RB, Freier JE and Lien JC (1983) Transovarial transmission of Dengue virus by mosquitoes: *Aedes albopictus* and *Aedes aegypti*. American Journal of Tropical Medicine and Hygiene 32: 1108–1119

Sabatini A, Rainieri V, Trovato G and Coluzzi M (1990) *Aedes albopictus* in Italia e possibile diffusione della specie nell'area mediterranea. Parassitologia 32(3): 301–304

Samanidou-Voyadjoglou A and Darsie R (1993) An annotated checklist and bibliography of the mosquitoes of Greece (Diptera: Culicidae). Mosquito Systematics 25(3): 177–185

Sardelis MR, Dohm DJ, Pagac B, Andre RA and Turell MJ (2002) Experimental transmission of Eastern Equine Encephalitis virus by *Ochlerotatus j. japonicus* (Diptera: Culicidae). Journal of Medical Entomology 39(3): 480–484

Schaffner F (2002) Rapport scientifique des opérations de surveillance et de traitement d'*Aedes albopictus* et autres espèces exotiques importées. Internal administrative report, EID Méditérranée – ADEGE, Montpellier, France, 31 pp

Schaffner F and Karch S (2000) Première observation d'*Aedes albopictus* (Skuse,1894) en France métropolitaine. Comptes Rendus de l'Académie des Sciences, Paris, Sciences de la Vie / Life Sciences, 323(4) : 373–375

Schaffner F, Bouletreau B, Guillet B, Guilloteau J and Karch S (2001) *Aedes albopictus* (Skuse, 1894) established in metropolitan France. European Mosquito Bulletin 9: 1–3

Schaffner F, Chouin S and Guilloteau J (2003) First record of *Ochlerotatus (Finlaya) japonicus japonicus* (Theobald, 1901) in metropolitan France. Journal of the American Mosquito Control Association 19(1): 1–5

Sprenger D and Wuithiranyagool T (1986) The discovery and distribution of *Aedes albopictus* in Harris County, Texas. Journal of the American Mosquito Control Association 2(2): 217–219

Torres Cañamares F (1979) Breve relación crítica de los mosquitos españoles. Revista de Sanidad e Higien Pública 53: 985–1002

Turell MJ, O'Guinn ML, Dohm DJ and Jones JW (2001) Vector competence of North American Mosquitoes (Diptera: Culicidae) for West Nile virus. Journal of Medical Entomology 38(2): 130–134

Urbanelli S, Bellini R, Carrieri M, Sallicandro P and Celli G (2000) Population structure of *Aedes albopictus* (Skuse): the mosquito which is colonizing Mediterranean countries. Heredity 84(3): 331–337

Vazeille-Falcoz M, Adhami J, Mousson L and Rodhain F (1999) *Aedes albopictus* from Albania: a potential vector of Dengue viruses. Journal of the American Mosquito Control Association 15(4): 475–478

Biological Invasions (2005) 7: 99–106

Holocene turnover of the French vertebrate fauna

Michel Pascal* & Olivier Lorvelec
*INRA, Station SCRIBE, Équipe Gestion des Populations Invasives, Campus de Beaulieu, 35042 Rennes Cedex, France; *Author for correspondence (e-mail: pascal@beaulieu.rennes.inra.fr; fax: +33-2-23485020)*

Received 4 June 2003; accepted in revised form 30 March 2004

Key words: biological invasion, extinction, France, spatial scale, temporal scale, vertebrate

Abstract

Comparing available paleontological, archaeological, historical, and former distributional data with current natural history and distributions demonstrated a turnover in the French vertebrate fauna during the Holocene (subdivided into seven sub-periods). To this end, a network of 53 specialists gleaned information from more than 1300 documents, the majority never cited before in the academic literature. The designation of 699 species as native, vanished, or non-indigenous in France or in one or more of its biogeographical entities during the Holocene period was investigated. Among these 699 species, 585 were found to belong to one or more of these categories. Among the 154 species that fit the definition of non-indigenous, 86 species were new species for France during the Holocene. Fifty-one that were autochthonous vanished from France during this period. Among these 51 species, 10 (two birds and eight mammals) are now globally extinct. During the last 11 millennia, the turnover in the French vertebrate fauna yielded a net gain of 35 species. On a taxon-by-taxon basis, there was a gain in the sizes of the ichthyofauna (19 : 27%), the avifauna (10 : 3%) and the herpetofauna (7 : 9%) and a loss in the mammalian fauna (−1 : 1%). Values of a per-century invasion index were less than 1 between 9200 BC and 1600 AD but increased dramatically after this date. An exponential model fits the trajectory of this index well, reaching the value of 132 invasions per century for the last sub-period, which encompasses 1945–2002. Currently, the local ecological and economic impacts of populations of 116 species (75% of the 154 that satisfied the criteria for non-indigenous) are undocumented, and the non-indigenous populations of 107 vertebrate species (69%) are unmanaged. The delay in assessing the ecological and economical impact of non-indigenous species, which is related to a lack of interest of French academic scientists in the Science and Action programmes, prevents the public from becoming informed and hinders the debates needed to construct a global strategy. For such a strategy to be effective, it will have to be elaborated at a more global scale than in just France – definitely at least in Europe.

Introduction

From the beginning of the second half of the 20th century, many scientists have recognized the detrimental effect that biological invasions may have on the species and functioning of invaded ecosystems; Elton (1958) was a notable early example. Nevertheless, it was only during the 1990 Rio conference that the subject arose in an international political forum. Since then, many documents have been produced covering many aspects of the problem. One major issue concerns the causes of the recent increase in the number of biological invasions. Among identified causes, probably the main one is the great increase in the volume of international trade after World War II, accelerated by an international easing of trade restrictions (Jenkins 1996; Mack et al. 2000 i.a.).

Despite its political and international standing, the subject was not discussed much in public and scientific forums in Europe during the 1990s; French governments, for example, never stressed the issue. What can account for this neglect? Taking France as an example, we can give two potential explanations (these may not be exhaustive). The first is structural, the second more cultural:

- The subject falls under the jurisdiction of many ministries (i.e., Ministry of the Environment, Ministry of Agriculture and Forestry, Ministry of Transport, Ministry of Foreign Affairs, Ministry of Health). Gathering the appropriate representatives of all these ministries in a single forum in order to formulate a general policy requires strong political leadership. The emergence of this issue implies that politicians clearly perceive the global risk posed by biological invasions. 'Global' in this sentence means that the risk surpasses the boundaries of traditional ministry jurisdiction and requires an assessment of the number, kind, and importance of impacts of non-indigenous species at a sufficiently large temporal and spatial scale.
- Many people, including scientists, perceive all European ecosystems, except for those few under protection, as having been influenced by humans for such a long time that they have lost their biological interest. This point of view implies a lack of concern for 'ordinary nature,' despite the fact that it comprises the major part of the country. A second point is that preventing introductions and managing populations of non-indigenous species appears to be impossible to many people. Moreover, when successful foreign examples of such prevention or management are presented, the response is often that local French circumstances lessen the relevance of foreign examples.

If accepted, these two explanations imply that pertinent data and knowledge about biological invasions in France are unavailable or uninterpretable for non-specialists. With the goal of clarifying perception of the issue, the French Ministry of the Environment ordered a synthesis of knowledge of the history, the biogeographical patterns, and impacts of invasions of France by vertebrates, as an exemplary taxon.

The aim of this paper is to summarize the reasoning and main results of this report (Pascal et al. 2003).

Methods

The area under study was restricted to the French European territory (including the Channel, West Atlantic, and French Mediterranean Islands), with the overseas departments (Martinique, Guadeloupe, and Reunion Islands, and French Guyana) and the French overseas territories excluded. As European France includes several biogeographical entities, it was divided into 11 biogeographical entities for terrestrial vertebrates (Tetrapoda) and six hydrogeographical basins for fishes. The 11 biogeographical entities include two insular ones (the Channel and Atlantic islands (1) and Corsica (2)), four mountainous ones (Alps (3), Pyrénées (4), Massif Central (5), Vosges, Jura and Ardennes (6)), three sets of plains (Atlantic plains, the northern limit is the Loire River and the southern limit is the Pyrénées (7), Paris Basin northward and eastward of the Seine River (8), Paris Basin southward of the Seine River, with Normandy and Brittany (9)), the French Mediterranean area (10) and the Rhone and Saone valleys, which were major ways of invasion (11). The six hydrogeographical basins are those defined by Persat and Keith (1997) and Keith (1998). Terrestrial and fresh and brackish water ecosystems were studied, but strictly marine ones were not.

The time interval considered was the Holocene period – 9200 BC to the present. This period starts with the end of the great modifications induced by the last climate warming. Among these modifications are the end of the last marine transgression and the stabilization of the west European vertebrate fauna, the majority of the cold-adapted species having migrated to the North, and the majority of the species of the Spanish, Italian and Dalmatian refugia having colonized at least the southern part of France. The Holocene was divided into seven sub-periods as follows:

1. 9200 BC–3000 BC: agriculture, animal-breeding, and the first villages appeared in western Europe.
2. 3000 BC–0: western European landscapes recorded the first strong anthropogenic influences of agriculture and animal-breeding. This sub-period ended with the *Pax Romana*.
3. 0–1600 AD: many events that strongly affected the landscape occurred, among them

being the Middle Age deforestation episode. This sub-period ended with the beginning of the global European Diaspora.

4. 1600–1800 AD: the European Diaspora was nearly complete and trade allowed many taxa to move between continents.
5. 1800–1914 AD: the advent of industry caused a dramatic evolution of the landscape because of new agricultural, silvicultural, and husbandry practices. The number of zoological gardens increased in Western Europe, promoting the introduction of non-indigenous species.
6. 1914–1945 AD: the two World Wars generated a large increase in the amount of international exchange and sped up the evolution of transport technology with substitution of a motor fleet for a sailing one, plus the development of roads, canals and railway networks.
7. 1945–2002 AD: during this half century, the western European landscape was dramatically modified by increasing urbanization, a rural exodus, and the further evolution of agriculture, silviculture, and husbandry. The perception of nature by citizens shifted as a consequence of the increasing fraction of the total population living in cities rather than farming communities. One consequence of this shift was the increasing number of non-indigenous pets.

We adopted a cladistic taxonomy, restricting ourselves to the species level for wild as well as for feral populations. The scientific nomenclatures used were those of Keith and Allardi (2001) for fishes, Gasc et al. (1997) for amphibians and reptiles, Dubois et al. (2000) for birds, and Wilson and Reeder (1993) for mammals.

For the purposes of this study, we defined a biological invasion as an event in which a species increased its distributional area during a specific period of time (whether or not because of human activities) and founded at least one self-perpetuating population in the newly invaded area.

This definition led to two corollaries:

– A species was regarded as autochthonous in France during the Holocene if it was believed to reproduce at the beginning of the Holocene in terrestrial, fresh, or brackish water ecosystems of at least one delineated biogeographical or hydrogeographical entity. If this species is allochthonous (non-indigenous) for one or more biogeographical or hydrogeographical entities, it satisfies the definition of biological invasion and was tallied as both autochthonous and allochthonous in France. At present, such a species may be present in France or it may be absent following a temporary disappearance. In the case of a temporary disappearance, the species accords with the definition of biological invasion and was tallied as autochthonous, disappeared, and allochthonous in France.

– A species was regarded as allochthonous in France during the Holocene period if it was believed not to reproduce at the beginning of the Holocene in the terrestrial, fresh, and brackish water ecosystems of France and is now represented by one or several self-sustaining populations. Self-sustaining means that the non-indigenous population does not require continuing recruitment from external sources in order to persist.

These three definitions share two features: history and *in situ* reproduction. Restriction to species with *in situ* reproduction resulted in discarding from the study all species that used French territories for various biological functions other than breeding. Among such species are birds using France for wintering or for migration, and amphihaline thalassotokous fishes. On the contrary, marine species such as sea turtles and seals were counted because they reproduce on the seashore. Lack of paleontological or archaeological data prevented us from adequately assessing the presence and reproduction at the beginning of the Holocene of many species belonging to the present French fauna; these are what Carlton (1996) calls 'cryptogenic' species, because their geographic origin is unknown. These species were considered as native unless we had proof to the contrary.

We used a simple typology of biological invasions. A first category included spontaneous biological invasions – that is, those whose arrival was not obviously related to human activities. The second category consisted of biological invasions that were inadvertently induced or facilitated by human activities but that were not transported by humans. The third category included both unintentional and intentional introductions transported by humans. This typology followed those elaborated by the Invasive Species Specialist Group (ISSG) of the IUCN (Anonymous 1999; Shine et al. 2000) and the

Invasive Species as defined by the ISSG were a subset of the third category.

For each tallied species, available paleontological, archaeological, and historical data were synthesized and compared to information on its natural history and past and present distributions. This synthesis was accomplished in order to classify the species as native, extinct, or vanished from France, or allochthonous based on all available information. For each non-indigenous species, we collected available data about the means and time of its arrival in France, its present numbers and distribution, and its impact on recipient ecosystems, including the role it may have played as a pathogen vector or reservoir. Data about management operations were also gathered wherever available.

Paleontological and archaeozoological information was extracted not only from the specialized literature, but also from the PTH database (PTH 1998; Vigne 1998) for mammals, the HAE-FAR (1993) database for birds, a Pleistocene synthesis (Mourer-Chauviré 1975; Vilette 1983; Laroulandie 2000; Louchart 2001) for both those taxa, and from Vigne et al. (1997), d'Hervet and Salotti (2000), and Bailon (2001) for the herpetofauna.

The present French fish distributions were gleaned from Keith (1998) and Keith and Allardi (2001). The present world, European, and French amphibian and reptile distributions were extracted from Anonymous (1987), Castanet and Guyetant (1989), Grossenbacher (1988), Hofer et al. (2001), Gasc et al. (1997), Parent (1981) and Delaugerre and Cheylan (1992), for birds from Voous (1960), Yeatman-Berthelot and Jarry (1994) and Dubois et al. (2000), and for mammals from Wilson and Reeder (1993), Mitchell-Jones et al. (1999) and Fayard (1984). Overall, the synthesis includes information from more than 1300 documents. Among them, some are academic papers, but the majority are reports, theses, or grey literature documents often issued during the last decade and never quoted before in academic literature. A network of 53 specialists amassed and verified all this information.

Results and discussion

We investigated the status of 699 species as native, vanished, or alien in France or one or several of its biogeographical entities during the Holocene period. Among these 699 species, 585 satisfied the definition of one or more of these categories. These 585 species include seven bird and three mammal species autochthonous in France that invaded the country after a temporary total disappearance and two allochthonous species that are presently absent after being present for several centuries.

Among these 585 species, 154, that is more than one-fourth, invaded France or at least one of its 11 biogeographical entities or six hydrogeographical basins during the Holocene period. Those species founded one or several populations that satisfied the criteria for a biological invasion.

Among these 154 species, 86, that is more than half, were species new to France during the Holocene if we consider France as a single geographical entity. Nevertheless, 68 species that were native in one or several French biogeographical entities invaded another entity during the Holocene. This result suggests that, as far as the topic of invasion is concerned, political or administrative entities are not adequate for an analysis of the subject.

Among the 585 species of the Holocene French vertebrate fauna, 51 autochthonous ones vanished during the Holocene (2 fishes, 1 amphibian, 1 reptile, 28 birds and 19 mammals). Among these 51 species, 10 (two birds and eight mammals) are now globally extinct.

A great discrepancy appears between the distribution of those 585 species among the four main vertebrate taxa and the sizes of those taxa in the world biota. With 308 species (53%), the French avifauna predominates, followed by the mammals (127 species; 22%), then the herpetofauna (80 species; 14%), and finally the ichthyofauna (70 species; 12%). This distribution differs strikingly from that of the world biota, in which herpetofauna are in the first place with 13,605 species (Frost 2002; Uetz et al. 2002) followed by avifauna (9968 species; Peterson 2002), freshwater ichthyofauna (9966 species among the 27,365 marine and freshwater species; Nelson 1994; Froese and Pauly 2003) and mammals (4629; Wilson and Reeder 1993).

The distribution of the 154 species that invaded the French biogeographical or hydrogeographical entities during the Holocene among the four main vertebrate taxa also shows a discrepancy. However, the ranking of taxa calculated for the total 585 species changes dramatically if, for each taxon, we tally the ratio between the number of species that invaded the French biogeographical entities and the total number of native and alien species. Thirty one (44%), 17 (21%), 68 (22%) and 38 (30%) species of the ichthyofauna, herpetofauna, avifauna, and mammalian fauna, respectively, are or were represented by one or several allochthonous populations during the Holocene in France.

This last ranking remains if this analysis is restricted to species that are allochthonous for the entire French territory: 21 (30%), 9 (11%), 38 (12%) and 18 (14%) species of the ichthyofauna, herpetofauna, avifauna, and mammalian fauna, respectively, are new in the French vertebrate fauna. Consequently, the sizes of the French ichthyofauna and mammalian fauna were more affected by biological invasions during the Holocene than were the avifauna and herpetofauna.

During the last 11 millennia, the turnover in the French vertebrate fauna led to an increase of 35 species (86–51), thus 6% of the total number (585). In the future, this figure may be revised downwards because several species, mainly among the birds, probably disappeared from France during the first nine Holocene millennia but were not counted in this category, as their disappearance is not yet documented.

If we examine the results of the turnover taxon by taxon, we find an increase in the ichthyofauna (19 : 27%), the avifauna (10 : 3%), and the herpetofauna (7 : 9%) and a decrease in the mammalian fauna (−1 : 1%). If French vertebrates are typical, this result suggests the following hypothesis: the older the divergence date of a taxon, the greater is its capability to persist owing to a low rate of species disappearance and a high rate of successful invasion.

The number of vertebrate species that invaded France or one or several of its biogeographical entities during the seven Holocene sub-periods is shown in Figure 1. Except for a small Neolithic wave of invasion, the striking feature of this invasion process is, unsurprisingly, an acceleration at the beginning of the 19th century, with the number of invasions between 1945 and 2002 comprising 49% of the total.

This analysis does not account for variation among sub-periods in length. To compensate for this variation, we defined a century invasion index as the number of vertebrate species that invaded France or one or several of its biogeographical entities per century during each of the seven Holocene sub-periods (Figure 2). This

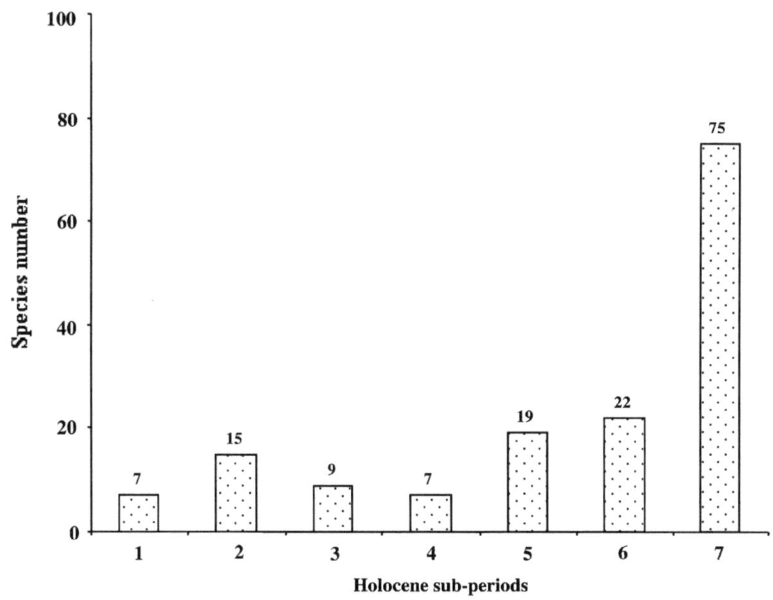

Figure 1. Number of vertebrate species that invaded France or one or several of its biogeographic entities during the seven Holocene sub-periods.

index value was less than 1 until 1600 AD, then increased dramatically. Its trajectory is well-fitted by an exponential model. It reaches the value of 132 invasions per century for the last sub-period, which encompasses 1945–2002. This last value may be overestimated because some of the new arrivals probably will not persist on French territory. The acceleration of the process may be also overestimated because past invasions are less well-documented than recent ones. Nevertheless, those two potential biases cannot by themselves explain the curve in Figure 2.

We compiled a list of all studies devoted to the assessment of impact for each species that invaded France or one or several of its biogeographical entities. We divided this dataset into four categories: no assessment, assessment of ecological impacts only, assessment of economic impacts only, assessment of both ecological and economic impacts. Studies devoted to ecological, economic, or both impact assessments were conducted for 11, 14, and 13 species, respectively. Consequently, at present, the ecological and economic impacts of 116 species (75% of 154) that are represented by allochthonous populations in France are completely undocumented.

We similarly compiled studies on management of allochthonous populations. Again, we constructed four categories: no management, management to reduce ecological impacts, management to reduce economic impacts, management to reduce both ecological and economic impacts. Allochthonous populations of 45 species are managed to reduce economic impacts and none are managed only for ecological purposes. Only *Rattus* populations (*R. rattus* and *R. norvegicus*) are managed for both economic (urban ecosystems) and ecological (insular ecosystems) purposes. Consequently, the allochthonous populations of 107 vertebrate species (69%) are currently not managed at all.

Conclusions

If the French vertebrate fauna is typical, these results show that national or administrative entities are not adequate to investigate the subject of biological invasions. Consequently, this subject must be investigated at the level of biogeographic entities.

The large discrepancy in the turnover among different vertebrate taxa shows that patterns for one taxon cannot always be generalized to others.

A sound understanding of the pace of biological invasions must account for time, not only to determine whether or not a species is allochthonous in a precise area, but also to grasp the changes in the tempo of the phenomenon. For this reason, paleontological, archeozoological, and historical data are of major interest despite difficulties in assessing their validity for comparison with recent data.

If allochthony alone does not justify managing populations, assessing the ecological and eco-

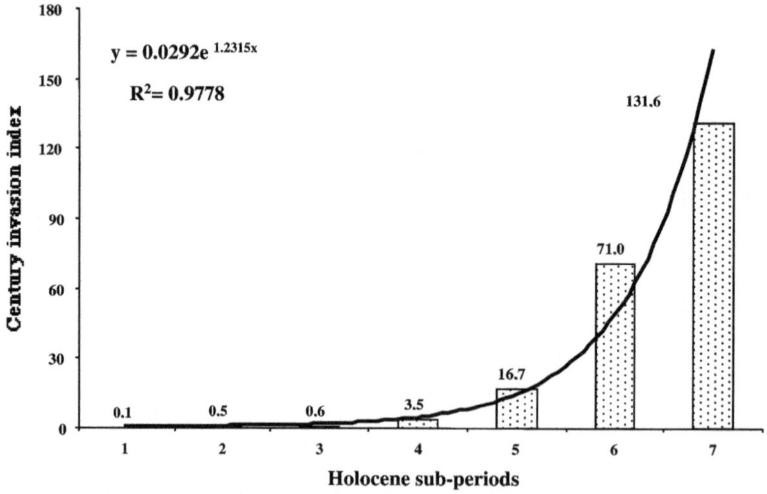

Figure 2. The century invasion index during the seven Holocene sub-periods.

nomic impact of non-indigenous species is a necessary precursor to development of an overall strategy. The exponential increase of the century invasion index, which reaches the value of 132 vertebrate invasions per century for the last 57 years, shows that France is not immune to the global flood of invasions. Further, the typical delay in assessing ecological and economic impacts of alien species hinders the dissemination of information to the public and thus delays the debates that must precede the development of an overall strategy.

This delay must be related to a lack of interest shown by French scientists in the Science and Action programmes. Although fundamental approaches to dealing with biodiversity and its recent evolution are promoted at the level of academic institutions (Anonymous 1995) and agencies (Fridlansky and Mounoulou 1996), no entity was devoted specifically to active management of and research on introduced species. As a consequence, for example, following two Environment Ministry calls for research projects on Biological Invasions (2000 and 2001), 30 projects were selected (Anonymous 2003), only four quoted the word 'management' in the title and two more the word 'control'.

Finally, to be efficient, such a strategy will have to be elaborated in Europe at least, as was proposed in the Convention on the conservation of European wildlife and natural habitats (Genovesi and Shine 2003).

Acknowledgements

We are grateful to Dan Simberloff who edited the English of this text and suggested several additions that clarified it. This research had financial support from the 'Sous-Direction de la Chasse, de la Faune et de la Flore Sauvage, Direction de la Nature et des Paysages' of the French Ministry of Environment.

References

Anonymous (1987) Atlas préliminaire des reptiles et amphibiens de France. Société Herpétologique de France, Paris

Anonymous (1995) Biodiversité et environnement. Rapport de l'Académie des Sciences No. 33. Technique et documentation. Lavoisier, Paris

Anonymous (1999) IUCN Guidelines for prevention of biodiversity loss due to biological invasion. Species 31–32: 28–42

Anonymous (2003) Programme de recherche invasions biologiques. Séminaire de programme Mars 2003. Ministère de l'Écologie et du Développement Durable, Paris

Bailon S (2001) Données fossiles des amphibiens et squamates de Corse: état actuel de la question. Bull Soc Sci Hist Nat Corse 696–697: 165–185

Carlton JT (1996) Biological invasions and cryptogenic species. Ecology 77(6): 1563–1655

Castanet J and Guyetant R (eds) (1989) Atlas de répartition des amphibiens et reptiles de France. Société Herpétologique de France, Paris

Delaugerre M and Cheylan M (1992) Atlas de répartition des batraciens et reptiles de Corse. Parc Naturel de Corse, École Pratique des Hautes Études, Bastia, France

Dubois PJ, Le Maréchal P, Olioso G and Yésou P (2000) Inventaire des oiseaux de France. Nathan, Paris

Elton CS (1958) The Ecology of Invasions by Animals and Plants. Methuen, London

Fayard A (ed) (1984) Atlas des mammifères de France. SFEPM, Muséum National d'Histoire Naturelle, Paris

Fridlansky F and Mounoulou J-C (1996) Programme national dynamique de la biodiversité et environnement? CNRS, Paris

Froese R and Pauly D (eds) (2003) FishBase. World Wide Web electronic publication (version 16 June 2003). www.fishbase.org

Frost DR (ed) (2002) Amphibian Species of the World: an Online Reference V2.21 (15 July 2002). http://research.amnh.org/herpetology/amphibia/index.html. American Museum of Natural History

Gasc J-P, Cabela A, Crnobrnja-Isailovic J, Dolmen D, Grossenbacher K, Haffner P, Lescure J, Martens H, Martinez Rica JP, Maurin H, Oliveira ME, Sofianidou TS, Veith M and Zwiderwijk A (eds) (1997) Atlas of Amphibians and Reptiles in Europe. Societas Europaea Herpetologica, Muséum National d'Histoire Naturelle (IEGB/SPN), Paris

Genovesi P and Shine C (2003) European Strategy on Invasive Alien Species. T-PVS (2002) 8 revised. Convention on the Conservation of European Wildlife and Natural Habitats. Conseil de l'Europe, Strasbourg, France

Grossenbacher K (1988) Atlas de distribution des amphibiens de Suisse. Documenta Faunistica Helvetiae, Ligue Suisse pour la Protection de la Nature, Centre Suisse de Cartographie de la Faune, Basel, Switzerland

HAE-FAR (1993) Database 'Hommes et animaux en Europe', gathered by F. Audoin-Rouzeau

Hervet S and Salotti M (2000) Les tortues pléistocènes de Castiglione (Oletta, Haute-Corse) et la preuve de leur indigénat en Corse. Comptes Rendus de l'Académie des Sciences de Paris, Série Sciences de la Terre et des Planètes 330: 645–651

Hofer U, Monney J-C and Duej G (2001) Les reptiles de Suisse. Répartition. Habitats. Protection. Koordinationsstelle für Amphibien und Reptilienschutz in der Schweiz (KARCH), Centre Suisse de Cartographie de la Faune (CSCF), Basel, Switzerland

Jenkins P (1996) Free trade and exotic species introductions. In: Sandlund OT, Schei PJ and Viken A (eds) Proceedings, Norway/UN Conference on Alien Species, pp 145–147. Directorate of Nature Management and

Norwegian Institut for Nature Research, Trondheim, Norway
Keith Ph (1998) Évolution des peuplements ichtyologiqucs dc France et stratégies de conservation. Thesis of the Rennes I University, Biologie, Rennes, France, no. 1997
Keith P and Allardi J (eds) (2001) Atlas des poissons d'eau douce de France. Patrimoines naturels, Muséum National d'Histoire Naturelle, Paris
Laroulandie V (2000) Taphonomie et achéozoologie des oiseaux en grotte: application aux sites paléolithiques du Bois Ragot (Vienne), de Combe Saunière (Dordogne) et de La Vache (Arière). Thesis of the Bordeaux I University, Bordeaux, France, no. 2341
Louchart A (2001) Les oiseaux du Pléistocène de Corse et données concernant la Sardaigne. Bulletin de la Société de Science et d'Histoire Naturelle de Corse 696–697: 187–221
Mack RN, Simberloff D, Lonsdale WM, Evans H, Clout M and Bazzaz FA (2000) Biotic invasions: causses, epidemiology, global consequences, and control. Ecological Applications 10(3): 689–710
Mitchell-Jones AJ, Amori G, Bogdanowicz W, Krystufek B, Reijnders PJH, Spitzenberger F, Stubb M, Thissen JB, Vohralik V and Zima J (1999) The Atlas of European Mammals. Academic Press, London
Mourer-Chauviré C (1975) Les oiseaux du Pléistocène moyen et supérieur de France. Thesis of the Claude Bernard University, Lyon, France, no. 75–14
Nelson JS (1994) Fishes of the World, 3rd edn. John Wiley & Sons, New York
Parent GH (1981) Matériaux pour une herpétofaune de l'Europe occidentale. Contribution à la révision chorologique de l'herpétofaune de la France et du Benelux. Bulletin de la Société Linnéenne de Lyon 50(3): 86–111
Pascal M, Lorvelec O, Vigne J-D, Keith P and Clergeau P (eds) (2003) Évolution holocène de la faune de Vertébrés de France: invasions et extinctions. Institut National de la Recherche Agronomique, Centre National de la recherche Scientifique, Muséum National d'Histoire Naturelle, Ministère de l'Aménagement du Territoire et de l'Environnement (Direction de la Nature et des Paysages), Paris
Persat H and Keith P (1997) La répartition géographique des poissons d'eau douce de France, qui est autochtone et qui ne l'est pas? Bulletin Français de Pêche et Pisciculture 344–345: 15–32
Peterson AP (2002) Zoonomen Nomenclatural Data. http://www.zoonomen.net
PTH (1998) Database compiled between 1994 and 1998 for the National Programme on Biological Diversity (PNDB) of the National Centre of the Scientific Research (CNRS) for the specific project 'Processus Tardiglaciaires et Holocènes de mise en place des faunes actuelles' (PTH). Scientific management of the database: Archéozoologie et Histoire des Sociétés, CNRS and Muséum National d'Histoire Naturelle (ESA 8045), Paris
Shine C, Williams N and Gündling L (2000) Guide pour l'élaboration d'un cadre juridique et institutionnel relatif aux espèces exotiques et envahissantes. IUCN, Gland/Cambridge/Bonn, pp I–XVI + 1–164
Uetz P, Etzold T and Chenna R (eds) (2002) The European Molecular Biology Laboratory (EMLB) Reptile Database. Systematics Working Group of the German Herpetological Society (DGHT). http://www.embl-heidelberg.de/~uetz/LivingReptiles.html
Vigne J-D (1998) Processus de mise en place de la faune actuelle d'Europe occidentale. In: Dynamique de la Biodiversité et Environnement, pp 36–39. CNRS, Paris
Vigne J-D, Bailon S and Cuisin J (1997) Biostratigraphy of amphibians, reptiles, birds and mammals in corsica and the role of man in the Holocene Faunal Turnover. Anthropozoologica 25–26: 587–604
Vilette P (1983) Avifaunes du Pléistocène final et de l'Holocène dans le sud de la France et en Catalogne. Laboratoire de Préhistoire Paléthnologique, Atacina, Carcassonne, France
Voous KH (1960) Atlas of European Birds. Elsevier, Amsterdam
Wilson DE and Reeder DAM (eds) (1993) Mammals Species of the World: a Taxonomic and Geographic Reference. Smithsonian Institution Press. Washington, DC
Yeatman-Berthelot D and Jarry G (eds) (1994) Nouvel atlas des oiseaux nicheurs de France. 1985–1989. Société Ornithologique de France, Paris

Biological Invasions (2005) 7: 107–116

Life-history traits of invasive fish in small Mediterranean streams

Anna Vila-Gispert, Carles Alcaraz & Emili García-Berthou*
*Department of Environmental Sciences & Institute of Aquatic Ecology, University of Girona, 17071 Girona, Catalonia, Spain; *Author for correspondence (e-mail: emili.garcia@udg.es; fax: +34-972-418150)*

Received 4 June 2003; accepted in revised form 30 March 2004

Key words: Iberian Peninsula, invasive freshwater fish, life-history traits, river basin size, water flow regulation

Abstract

We compared the life-history traits of native and invasive fish species from Catalan streams in order to identify the characters of successful invasive fish species. Most of the exotic fish species were characterized by large size, long longevity, late maturity, high fecundity, few spawnings per year, and short reproductive span, whereas Iberian native species exhibited predominantly the opposite suite of traits. Species native to the southeastern Pyrenees watershed were also significantly different from species native to the rest of the Iberian Peninsula but not native to this watershed. Iberian exotic species come predominantly from large river basins, whereas Catalan streams (and other small, coastal river basins) correspond to basins and streams of a smaller size and different hydrology, with differences in species composition and life-history traits of fish. The occurrence and spread of invasive species was not significantly related to life-history traits but to introduction date. The successful prediction of future invasive species is limited due to small differences in life-history and ecological traits between native and exotic species. Fecundity, age at maturity, water quality flexibility, tolerance to pollution and habitat seem the most discriminating life-history variables.

Introduction

Stream fish faunas are increasingly subject to alteration by introduction of non-native fishes. Successful establishment of non-native forms varies widely between geographic regions (38–77%), but is generally greater in areas that are either altered by man or initially poor in fish species (Ross 1991). Mediterranean streams have strong seasonal patterns of flow: low flow in the hot summer drought and flash floods during autumn and spring storms. Interannual variability in precipitation is high while lengthy periods of drought are common. As a consequence, the native fish fauna is depauperate and highly endemic (Doadrio et al. 1991). The natural variability in environmental conditions of many Iberian streams has been greatly reduced by water regulation. In addition, industrial waste and sewage effluents cause water quality to deteriorate. These profound modifications to the fluvial systems of this region directly threaten the native fish fauna and favor the invasion of non-native species (Elvira 1995, 1998).

Introduced species have been successful in the Mediterranean-type climate of California in good part because the natural environment has been so altered. Free-flowing streams have increasingly been turned into reservoirs, regulated streams, and ditches (Moyle 1995). The failure of non-native species to become established despite repeated invasions is best attributed to their inability to adapt to the local hydrological regime (Minckley and Meffe 1987; Baltz and Moyle 1993). In contrast, native species are adapted to flooding regimes through a combination of life-history strategies and physiological tolerances (Moyle et al. 1986). The factors that determine whether a species will be an invader or not include both the species and the habitat (Williamson and Fitter 1996). Here we concentrate on the characters of the species, basically life-history and ecological traits. We compare native and invasive species from Catalan streams in order to identify the characters of successful invasive fish species.

Materials and methods

The southeastern Pyrenees watershed is located in the northeast of Spain, has a surface area of 16,826 km^2 and includes nine river basins that flow into the Mediterranean Sea (Aparicio et al. 2000). From the north to the south, the river basins are Muga, Fluvià, Ter, Tordera, Besòs, Llobregat, Foix, Gaià, and Francolí. This watershed, limited by a larger river (Ebro) and the Pyrenees mountains, constitutes a biogeographical unit on the basis of its freshwater fishes (Doadrio 1988) and a management unit for Spanish water authorities. The Ter and Llobregat rivers are the largest in terms of length and discharge. They have a main peak flow in spring caused by snow melt and rain, and a secondary one in autumn due to rainfall. The other basins are short because they rise in littoral mountains and their valley slopes are relatively steep down to their mouths. They have the highest flow in autumn and a strong flow reduction in summer. River flows have been modified considerably by the construction of 10 large dams, and countless small dams and water diversion barriers (Aparicio et al. 2000).

Estimates of fish life-history traits were obtained from our own data (Vila-Gispert and Moreno-Amich 2000) and from literature sources specified in Table 1 (see also Vila-Gispert et al. 2002). Species considered were those listed in Doadrio (2001), excluding *Anguilla anguilla* due to their peculiar reproductive biology. Species were classified as native to the south-eastern Pyrenees watershed, species native to the rest of the Iberian Peninsula but not native to this watershed (hereafter referred to as Iberian exotic), and species not native to the Iberian Peninsula (here after referred to as foreign exotic). The results not distinguishing between Iberian and foreign exotic were similar but less informative. In general, data about ecological variables were obtained from Doadrio (2001) and Oberdorff et al. (2002), date of introduction from Elvira and Almodóvar (2001), and distribution and occurrence of fish species from Aparicio et al. (2000) (Table 1). When no reliable data were found for a given variable, that cell in the species variable matrix was left blank and any calculations calling for the variable eliminated the species from the analysis. Whenever maturation and maximum length data were reported for the sexes separately, we used the estimates for females. In some instances, standard lengths (SL) were calculated from total lengths or fork lengths using published conversion equations.

Data for the following variables were used in the analyses: (1) maximum standard length reported in millimeters SL; (2) age at maturity (in months); (3) reproductive span (in months); (4) spawning type categorized as 1 (single spawning per year), 2 (from two to four spawnings per year), or 3 (more than four spawnings per year); (5) fecundity as the average number of vitellogenic oocytes of mature females in a single mature ovary or spawning event; (6) egg diameter (the average diameter of the largest oocytes in fully developed ovaries, to the nearest 0.01 mm); (7) longevity as the maximum age estimated (in years); (8) parental care following Winemiller (1989), quantified as $\sum x_i$ for $i = 1$ to 3 ($x_1 = 0$ if no special placement of zygotes, 1 if zygotes are placed in a special habitat (e.g. scattered on vegetation, or buried in gravel), and 2 if both zygotes and larvae are maintained in a nest; $x_2 = 0$ if no parental protection of zygotes and larvae, 1 if a brief period of protection by one sex (<1 month), 2 if a long period of protection by one sex (>1 month) or brief care by both sexes, and 4 if lengthy protection by both sexes; $x_3 = 0$ if no nutritional contribution to larvae (yolk sac material is not considered here), 2 if brief period of nutritional contribution to larvae (=brief gestation (<1 month) with nutritional contribution in viviparous forms), 4 if long period of nutritional contribution to larvae or embryos (=long gestation (1–2 months) with nutritional contribution), or 8 if extremely long gestation (>2 months)); (9) gregariousness referred to the adults was coded as a binary dummy variable (0 (no), 1 (yes); (10)

Table 1. Native, Iberian exotic, and foreign exotic fish species from the south-eastern Pyrenees watershed and reference sources of data.

Species	Code	Origin	References
Aphanius iberus	AIB	Native	Vargas and Sostoa (1997)
Barbus haasi	BHA	Native	Aparicio and Sostoa (1998)
Barbus meridionalis	BME	Native	Ouslé and Quignard (2001); Doadrio (2001)
Chondrostoma arcasii	CAR	Native	Rincón and Lobón-Cerviá (1989)
Gasterosteus gymnurus	GGY	Native	Winemiller and Rose (1992)
Salaria fluviatilis	SFL	Native	Viñolas (1986); Sostoa (1990); Vila-Gispert and Moreno-Amich (2000)
Salmo trutta	STR	Native	Lobón-Cerviá et al. (1986)
Squalius cephalus	SCE	Native	Casals (1985); Duslé and Quignard (2001)
Barbatula barbatula	BBA	Iberian	Doadrio (2001)
Barbus graellsii	BGR	Iberian	Sostoa (1990); Miñano et al. (2000); Doadrio (2001)
Chondrostoma miegii	CMI	Iberian	Chappaz et al. (1989); Miñano et al. (2000)
Cobitis calderoni	COC	Iberian	Doadrio (2001)
Cobitis paludica	CPA	Iberian	Lobón-Cerviá and Zabala (1984); Oliva-Paterna et al. (2002)
Cottus gobio	CGO	Iberian	Mann et al. (1984); Duslé and Quignard (2001)
Phoxinus phoxinus	PPH	Iberian	Mills and Eloranta (1985)
Alburnus alburnus	AAL	Foreign	Mackay and Mann (1969)
Ameiurus melas	AME	Foreign	Winemiller and Rose (1992)
Carassius auratus	CAU	Foreign	Lelek (1980); Penaz and Kokes (1981); Duslé and Quignard (2001); Doadrio (2001)
Cyprinus carpio	CCA	Foreign	Fernández-Delgado (1990)
Esox lucius	ELU	Foreign	Wright and Schoesmith (1988)
Gambusia holbrooki	GHO	Foreign	Fernández-Delgado (1989)
Gobio gobio	GGO	Foreign	Lobón-Cerviá et al. (1991)
Lepomis gibbosus	LGI	Foreign	Crivelli and Mestre (1988); Vila-Gispert and Moreno-Amich (2000); Copp et al. (2002)
Micropterus salmoides	MSA	Foreign	Winemiller and Rose (1992)
Perca fluviatilis	PFL	Foreign	Mann (1978)
Rutilus rutilus	RRU	Foreign	Vila-Gispert and Moreno-Amich (2000)
Salvelinus fontinalis	SFO	Foreign	Winemiller and Rose (1992); Doadrio (2001)
Sander lucioperca	SLU	Foreign	Lehtonen et al. (1996); Duslé and Quignard (2001)
Scardinius erythrophthalmus	SER	Foreign	Holcik (1967); Vila-Gispert and Moreno-Amich (2000)
Tinca tinca	TTI	Foreign	Pimpicka (1990)

habitat was used to indicate habitat preferences of the adults along the longitudinal river gradient and was coded as 1 (upper), 2 (upper–middle), and 3 (lower); (11) trophic level was coded as 1 (herbivore/omnivore), 2 (invertivore), and 3 (piscivore); (12) coefficient of water quality flexibility (low values = narrow range of acceptable water quality) after Verneaux (1981) (see also Oberdorff et al. 2002); (13) coefficient of habitat flexibility (low values = narrow range of acceptable habitats) after Grandmottet (1983); and (14) tolerance coded as low (1) when species have a coefficient of water quality flexibility < 6 and a coefficient of habitat flexibility ≤ 0.1, medium (2) when species have a coefficient of water quality flexibility $= 6$ or <7 and a coefficient of habitat flexibility > 0.1 and <0.3, and high (3) when species have a coefficient of water quality flexibility ≥ 7 and a coefficient of habitat flexibility ≥ 0.3.

Bivariate relationships were analyzed using Pearson's correlation coefficient for numerical variables. Differences among species origin (native, Iberian exotic and foreign exotic) in the mean of numerical variables were tested for significance using a one-way analysis of variance (ANOVA). Post-hoc comparisons of means were made with Games–Howell (GH) test. Maximum length, fecundity, age at maturity and spawning type were $\log_{10}$ transformed for all analyses to improve linearity and homoscedasticity assumptions. For categorical variables such as gregariousness, tolerance, trophic level, and habitat, we used tests of independence (G-statistic) to test differences among species origin.

To explore patterns of association among variables and ordinate the species, principal component analysis (PCA) was performed on the 13 variables of the dataset containing 30 native and introduced fish species from the southeastern. Pyrenees watershed (Table 1). Kaiser–Meyer–Olkin's measure of sampling adequacy was used to assess the usefulness of a PCA. KMO ranges from 0 to 1 and should be well above 0.5 if variables are very interdependent and a PCA is useful.

To test relationships between life-history and ecological traits and species origin, we performed discriminant function analysis (DFA) based on 11 variables. DFA derives canonical variables from the set of variables in a manner that maximizes multiple correlations of the original variables within groups. Stepwise DFA based on 11 variables was also performed. Correlations between the first two DFA axes and variables associated with invasive success (river length occupied in kilometers, proportional occurrence, percentage change in river length occupied, and date of introduction) were analyzed using Pearson's correlation coefficient.

Because fecundity, age at maturity, egg diameter, and longevity are very dependent on fish size, differences among species origin subsets adjusted for maximum length were tested with analysis of covariance (ANCOVA). Homogeneity of regression coefficients (slopes) of the dependent-covariate relationship was tested with an ANCOVA design with the covariate-by-factor interaction. If the covariate-by-factor interaction was not significant (homogeneity of slopes), a standard ANCOVA was used to test significant differences in the *y*-intercept among populations (García-Berthou and Moreno-Amich 1993). All statistical analyses were performed with the SPSS for Windows 11.0.

Results

Correlations between all pairwise combinations of life-history and ecological traits are given in Table 2. Most of the variables were correlated and the Kaiser–Meyer–Olkin's measure of sampling adequacy (0.55) indicated the usefulness of a PCA, which explained 53% of the variation with two axes (Figure 1). As also seen with the factor loadings, the highest correlations were found between longevity, fecundity, maximum length, and age at maturity, which were positively intercorrelated and negatively related to

Table 2. Correlation matrix of life-history and ecological traits (Pearson's correlation).

Variables	2	3	4	5	6	7	8	9	10	11	12	13
1. Age at maturity	−0.60*	0.59*	0.73*	0.58*	−0.10	−0.69*	0.21	−0.23	−0.14	0.01	0.09	−0.10
2. Spawning type	–	−0.45**	−0.59*	−0.64*	0.05	0.71*	−0.29	0.18	−0.15	−0.02	−0.41	−0.08
3. Fecundity		–	0.74*	0.82*	−0.04	−0.69*	0.21	−0.23	0.28	−0.14	0.54**	0.28
4. Maximum length			–	0.79*	−0.20	−0.60*	0.40**	−0.12	−0.01	0.15	0.37	0.05
5. Longevity				–	−0.23	−0.48**	−0.05	−0.08	0.29	0.18	0.41	0.14
6. Gregariousness					–	0.08	0.01	−0.36	0.21	−0.49*	0.01	0.20
7. Reproductive span						–	−0.27	0.38**	0.12	−0.09	−0.30	−0.10
8. Egg diameter							–	−0.09	−0.46**	0.17	−0.48	−0.51**
9. Parental care								–	0.20	0.29	−0.16	−0.17
10. Habitat									–	−0.15	0.64*	0.37
11. Trophic level										–	−0.35	−0.20
12. Coefficient of water quality flexibility											–	0.66*
13. Coefficient of habitat quality flexibility												–

Correlations were based on all available data for south-eastern Pyrenees watershed fish species (*$P < 0.05$; **$P < 0.01$).

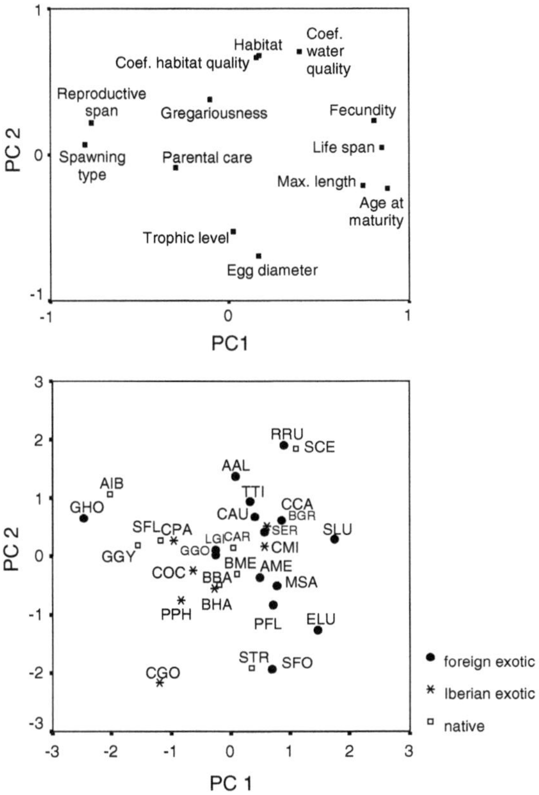

Figure 1. Principal component analysis of 13 life-history and ecological variables for native, Iberian exotic, and foreign exotic fish species from the southeastern Pyrenees watershed: (top) factor loadings of the variables, (bottom) species scores on the first two principal component axes. Species codes are given in Table 1. Symbols identify species origin.

reproductive span and spawning type. The first PCA axis identified a dominant gradient of life-history traits that contrasts species with large size, long longevity, late maturity, high fecundity, few spawning bouts per year, and short reproductive span (such as zander *Sander lucioperca*, largemouth bass *Micropterus salmoides*, and northern pike *Esox lucius*) with small species with short longevity, early maturity, low fecundity, multiple spawning bouts per year, and long reproductive span (such as mosquitofish *Gambusia holbrooki*, the Iberian toothcarp *Aphanius iberus*, and the freshwater blenny *Salaria fluviatilis*) (Figure 1). The second axis contrasts species with higher water and habitat quality flexibilities, small eggs, and of lower reaches on one end (mostly cyprinids) with species with low quality flexibilities, larger eggs, and of upper reaches on the other (e.g., brown trout, brook trout, and bullhead *Cottus gobio*).

Species origin affiliations are slightly apparent in the general pattern of ordination of species within regions in the plot of species scores on the first two PC axes (Figure 1). Foreign introduced species tended to score higher on PC1 than native and Iberian exotic species, whereas there are no clear differences on PC2 scores according to species origin. Most of foreign introduced species (such as zander, northern pike, largemouth bass, perch *Perca fluviatilis*, common carp *Cyprinus carpio*) combine large size, long longevity, late maturity, high fecundity, few spawning bouts per year, and short reproductive span. In contrast, most of the species with lowest PC1 (except mosquitofish), i.e. with the smallest size, shortest longevity, and lower fecundity among other features were native (such as the Iberian toothcarp, the blenny, and the three-spine stickleback *Gasterosteus gymnurus*) and the Iberian exotic species were intermediate.

Univariate comparisons in life-history and ecological traits among species origins revelead significant differences between native, Iberian exotic and foreign exotic species only in mean fecundity and coefficient of water quality flexibility (Table 3, Figure 2). A multivariate ANOVA was also significant (Wilks' lambda, $P = 0.036$). Post-hoc comparisons of means among species origin subsets identified significant differences in fecundity between native and foreign exotic species (GH test, $P = 0.027$) and between Iberian exotic and foreign exotic species (GH test, $P = 0.024$) but not between native and Iberian exotic. For categorical variables, tolerance and habitat frequencies depended on species origin (Table 4). Overall, 93% of the foreign exotic species were from lower habitats, whereas only 50% of the native and 43% of the Iberian exotic were found in lower habitats. Intermediate tolerance to pollution was displayed by 73% of the foreign exotic species but none of the native and Iberian exotic species.

DFA using species origin as the categorical response variable (Figure 3) confirmed this pattern of high fecundity (within-group correlation with the first DFA function = 0.38), long longevity (0.33), large size (0.31), lower habitat reaches (0.26), and few spawning bouts per year (−0.25) in association with foreign exotic species vs low fecundity, short longevity, small size, upper and middle habitat reaches, and multiple spawning

Table 3. Analyses of variance of life-history and ecological variables with species origin (native, Iberian exotic and foreign exotic): *F*-statistics, degrees of freedom (df), and *P*-values.

Variable	*F*	df	*P*
Longevity	3.05	2, 24	0.066
Fecundity	6.20	2, 27	0.006
Maximum length	3.14	2, 27	0.059
Spawning type	1.45	2, 26	0.291
Egg diameter	0.07	2, 26	0.932
Parental care	0.59	2, 27	0.559
Reproductive span	0.42	2, 27	0.660
Habitat	3.23	2, 27	0.055
Trophic level	0.34	2, 27	0.714
Coefficient of water quality flexibility	6.04	2, 12	0.015
Coefficient of habitat quality flexibility	0.79	2, 12	0.475

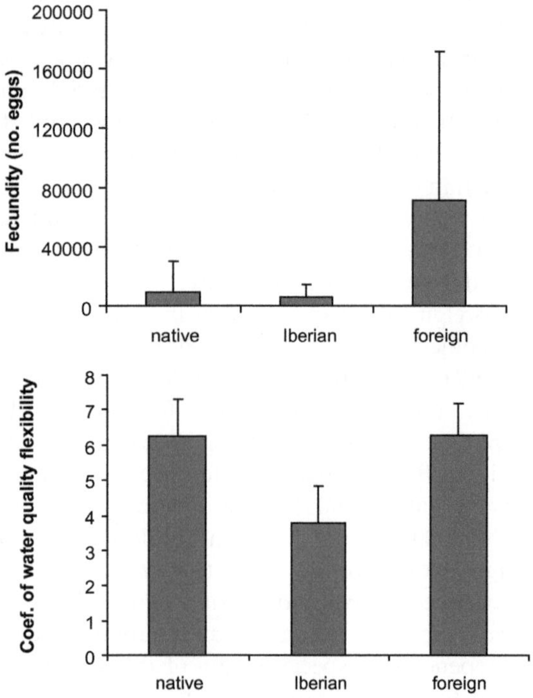

Figure 2. Means (+1 SE) of fecundity and coefficient of water quality flexibility by species origin (native, Iberian exotic and foreign exotic).

bouts in association with native and Iberian exotic species. The second discriminant axis of DFA separated Iberian exotic species with late maturity (0.42) from native species with early maturity. DFA correctly predicted the origin status for 83% of the species. Stepwise DFA selected that fecundity was the most discriminating variable and was enough to significantly separate native, Iberian exotic and foreign exotic groups of species (Wilks' $\lambda = 0.75$; $F = 3.9$; df = 2, 23; $P = 0.03$).

Table 4. Tests of independence of categorical variables with species origin (native, Iberian exotic and foreign exotic): *G*-statistics, degrees of freedom (df), and *P*-values.

Variables	*G*	df	*P*
Tolerance	9.76	4	0.045
Habitat	9.92	4	0.042
Gregariousness	0.58	2	0.748
Trophic level	7.47	4	0.111

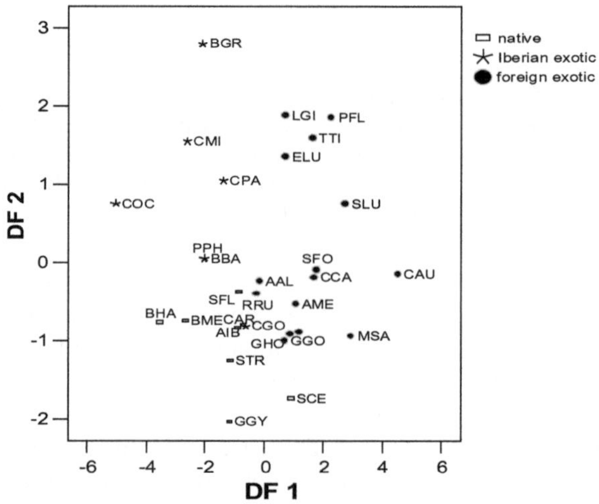

Figure 3. Species scores on the first two discriminant function (DFA) axes by species origin based on 13 life-history and ecological variables. Species code as in Table 1.

ANCOVA analyses showed that the slopes of the relationships between fecundity, age at maturity, egg diameter, and longevity with maximum length (covariate) did not significantly vary among species origins. The intercepts or adjusted means varied significantly ($F = 3.84$; df = 23, 2; $P = 0.04$) only for age at maturity (Figure 4).

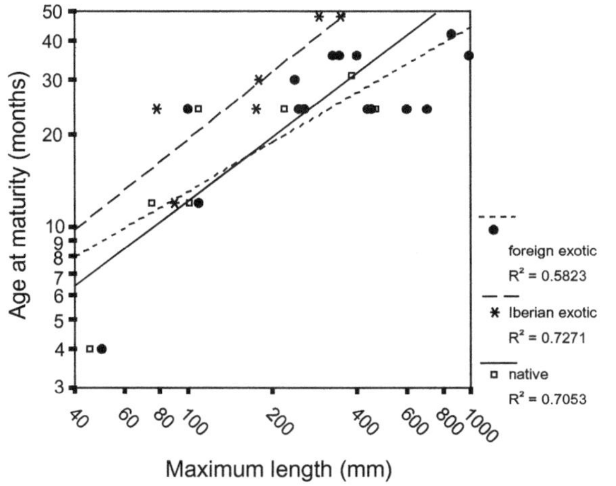

Figure 4. Relationship of age at maturity with the maximum fish size by species origin affiliations (native, Iberian exotic and foreign exotic). The age at maturity and maximum length were $\log_{10}$ transformed.

For a given length, Iberian exotic fish have a higher age at maturity.

Pearson's correlations between the first two DFA axes and variables associated with invasive success (river length occupied in km, proportional occurrence, percentage change in river length occupied, and date of introduction) are summarized in Table 5. The two discriminant functions were not correlated with any invasive success variable, whereas the proportional occurrence of species and river length occupied were negatively correlated with the date of introduction.

Discussion

Differences between native and exotic species characteristics were already apparent in the general pattern of ordination of species in the plot of species scores on the first two PC axes (Figure 1). Most of the foreign exotic species were located near one endpoint of the first axis of PCA, having large size, long longevity, late maturity, high fecundity, few spawnings per year, and short reproductive span, whereas Iberian exotic and native species exhibited predominantly the opposite suite of traits. The suite of life-history traits of foreign exotic species corresponds well to periodic life-history strategy defined by Winemiller (1989) and Winemiller and Rose (1992) which maximizes age-specific fecundity at the expense of optimizing turn-over time (turn-over times are lengthened by delayed maturation) and juvenile survivorship (see also Vila-Gispert and Moreno-Amich 2000, Vila-Gispert et al. 2002). Several authors (Cohen 1967; Boyce 1979; Baltz 1984) predict maximization of fecundity in response to predictable, seasonal environment variation. Environmental conditions of most Catalan streams have been altered by both pollution and the construction of dams which reduces strong flow variability and stabilizes downstream flows. As a consequence, most of the foreign exotic species that have successfully invaded Catalan streams come from seasonal habitats (central European and southeastern North American streams) that are more hydrologically stable. In contrast, native and Iberian exotic species were characterized by small size, short longevity, early maturity, low fecundity, multiple spawnings per year, and long reproductive span. Relative to foreign exotic species, native and Iberian exotic species displayed a more opportunistic life-history strategy (Winemiller 1989; Winemiller and Rose 1992) which enhances the intrinsic rate of population increase and consequently, the fitness of individuals in populations that frequently

Table 5. Correlation matrix (Pearson's correlation) of the first two axes of DFA and variables associated with invasive success (river length occupied in kilometers, proportional occurrence, percentage change in river length occupied, and date of introduction).

Variable	3	4	5	6
1. DFA 1	−0.01	−0.18	0.08	0.19
2. DFA 2	−0.03	0.08	0.10	−0.45
3. River length	–	0.90**	−0.14	−0.96**
4. Proportional occurrence		–	−0.24	−0.98**
5. Percentage change in river length			–	0.34
6. Date of introduction				–

River length was log-transformed (*$P < 0.05$; **$P < 0.01$).

colonize habitats over small spatial scales following disturbances.

DFA showed a trend of higher fecundity, longer longevity, larger size, lower reaches and fewer spawning bouts in association with foreign exotic species. In contrast, lower fecundity, shorter longevity, smaller size, upper reaches, and more spawning bouts were associated with Iberian exotic and native species. The highest proportion of native and Iberian exotic fishes in Catalan streams are found in headwaters and upper reaches of streams (Aparicio et al. 2000; Doadrio 2001), whereas most of the foreign exotic species are found in the middle and lower reaches. As pointed out by Minckley and Meffe (1987) and Moyle (1995), natural flooding provides the most evidence for slowing or precluding establishment of foreign exotic species. Headwater streams and upper reaches of Catalan streams experience strong seasonal patterns of flow: low flows in summer that restrict aquatic habitats to small isolated pools, and high flows in winter and spring. These environmental conditions prevent the invasion of large foreign exotic species more adapted to lentic habitats, whereas native and Iberian exotic species more adapted to strong seasonal patterns of flow could survive due to its lower sizes and to multiple spawnings that prevent the loss of all offspring after flood disturbances. In contrast, foreign exotic species successfully invade middle and lower reaches of Catalan streams where the effects of flow regulation are stronger. DFA also separated Iberian exotic species with late maturity from native ones with early maturity. Iberian exotic species come from large streams where the impacts of floods were lesser than in small Catalan streams. As a result, early maturity in species from Catalan streams maximizes the intrinsic rate of population increase after flood disturbances.

Invasive success measured as the river length occupied in kilometers, proportional occurrence, and percentage change in river length occupied was not correlated with any discriminant function axes, so it seemed that life-history and ecological traits could not be used to explain differences in invasive success. In contrast, the proportional occurrence of species and the river length occupied were negatively correlated with date of introduction of species, suggesting that introduction date is an important explanatory variable and that species more recently introduced will extend its distribution.

Surprisingly, native species were similarly distant in life-history traits from Iberian than from foreign exotic species (Figures 3 and 4). Iberian exotic species come predominantly from large river basins such as Guadalquivir (602 km), Guadiana (778 km), Duero (775 km), Tajo (940 km), and Ebro (928 km), whereas Catalan streams (excluding Ebro river) correspond to basins and streams of smaller size and different hydrology, due to the short distance from the mountains to the coastline (maximum length 167 km). These spatial-scale differences produce differences in species composition and life-history traits between fish from Catalonia (or other small, coastal basins) and fish from large Iberian rivers. As a result, it is as important to prevent the introduction of exotic fish species to the Iberian peninsula so as to prevent the translocation of Iberian species in basins where they are not native. Moreover, the identification of invasive fish features seems to profoundly depend on river basin size, as many other ecological patterns and processes.

Finally, we conclude that the possibility of prediction of success for future invasive species is limited due to small differences in life-history and ecological traits between native and exotic species. In our study, only fecundity, age at maturity, water quality flexibility, tolerance to pollution and habitat showed some significant differences. A multivariate analysis was more powerful in detecting such differences. Further studies in different regions and scales are needed to understand the role of life-history traits in invasive fish success.

Acknowledgements

This work was financially supported by the Spanish Ministry of Science and Technology (grant no. REN2003-00477/GLO). CA held a doctoral fellowship (FPU) from the Spanish Ministry of Education.

References

Aparicio E and de Sostoa A (1998) Reproduction and growth of *Barbus haasi* in a small stream in the N.E. of the Iberian peninsula. Archiv für Hydrobiologie 142: 95–110

Aparicio E, Vargas MJ, Olmo JM and de Sostoa A (2000) Decline of native freshwater fishes in a Mediterranean watershed on the Iberian Peninsula: a quantitative assessment. Environmental Biology of Fishes 59: 11–19

Baltz DM (1984) Life history variation among female surfperches (Perciformes: Embiotocidae). Environmental Biology of Fishes 10: 159–171

Baltz DM and Moyle PB (1993) Invasion resistance of an assemblage of native California stream fishes. Ecological Applications 3: 140–255

Boyce MS (1979) Seasonality and patterns of natural selection for life histories. American Naturalist 114: 569–583

Bruslé J and Quignard JP (2001) Biologie des poissons d'eau douce Européens. Techinque & Documentation, Paris

Casals F (1985) Biología i ecologia de *Leuciscus cephalus cephalus* (L., 1758) en el riu Matarranya. PhD Thesis, Department of Zoology, University of Barcelona, 171 pp

Chappaz R, Brun G and Olivari G (1989) Données nouvelles sur la biologie et l'écologie d'un poisson Cyprinidé peu étudié *Chondrostoma toxostoma* (Vallot, 1836). Comparison avec *Chondrostoma nasus* (L., 1766). Comptes Rendus de l'Académie des Sciences, Paris 309: 181–186

Cohen D (1967) Optimizing reproduction in a randomly varying environment when a correlation may exist between the conditions at the time a choice has to be made and the subsequent outcome. Journal of Theoretical Biology 16: 1–14

Copp GH, Fox MG and Kóvac V (2002) Growth, morphology and life-history traits of a cool-water European population of pumpkinseed *Lepomis gibbosus*. Archiv für Hydrobiologie 155: 585–614

Crivelli AJ and Mestre D (1988) Life history traits od pumpkinseed, *Lepomis gibbosus*, introduced into the Camargue, a Mediterranean wetland. Archiv für Hydrobiologie 111: 449–466

Doadrio I (1988) Delimitation of areas in tha Iberian Peninsula on the basis of frehwater fishes. Bonner Zoologische Beiträge 39: 113–128

Doadrio I (2001) Atlas y libro rojo de los peces continentales de España. Dirección General de Conservación de la Naturaleza, Madrid, 364 pp

Doadrio I, Elvira B and Bernat Y (1991) Peces continentales españoles. Inventario y clasificación de zones fluviales, ICONA, Madrid, 221 pp

Elvira B (1995) Conservation status of endemic freshwater fish in Spain. Biological Conservation 72: 129–136

Elvira B (1998) Impact of introduced fish on the native freshwater fish fauna of Spain. In: Cowx IG (ed) Stocking and Introduction of Fish, pp 186–190. Fishing New Books, Oxford

Elvira B and Almodóvar A (2001) Freshwater fish introductions in Spain: facts and figures at the beginning of the 21st century. Journal of Fish Biology 59: 323–331

Fernández-Delgado C (1989) Life-history patterns of the mosquito-fish, *Gambusia affinis*, in the estuary of the Guadalquivir River of south-west Spain. Freshwater Biology 22: 395–404

Fernández-Delgado C (1990) Life history patterns of the common carp, *Cyprinus carpio*, in the estuary of the Guadalquivir River in south-west Spain. Hydrobiologia 206: 19–28

García-Berthou E and Moreno-Amich R (1993) Multivariate analysis of covariance in morphometric studies of the reproductive cycle. Canadian Journal of Fisheries and Aquatic Sciences 50: 1394–1399

Grandmottet JP (1983) Principales exigencies des téléostéens dulcicoles vis à vis de l'habitat aquatique. Annales Scientifiques de l'Université de Franche-Comté 4: 3–32

Holcik J (1967) Life history of the rudd- *Scardinius erythrophthalmus* (Linnaeus, 1758) in the Klícava Valley Reservoir. Acta Societatis Zoologicae Bohemoslovenicae 31: 335–348

Lehtonen H, Hansson S and Winkler H (1996) Biology and exploitation of pikeperch, *Stizostedion lucioperca* (L.), in the Baltic Sea area. Annales Zoologici Fennici 33: 525–535

Lelek A (1980) Threatened Freshwater Fishes of Europe. Council of Europe. Strasbourg, 269 pp

Lobón-Cerviá J and Zabala A (1984) Observations on the reproduction of *Cobitis paludicola* de Buen, 1930 in the Jarama River. Cybium 8: 63–68

Lobón-Cerviá J, Montañés C and Sostoa A de (1986) Reproductive ecology and growth of a population of brown trout (*Salmo trutta* L.) in an aquifer-fed stream of Old Castille (Spain). Hydrobiologia 135: 81–94

Lobón-Cerviá J, Montañés C and Sostoa A de (1991) Influence of environment upon the life history of gudgeon, *Gobio gobio* (L.): a recent and successful colonizer of the Iberian Peninsula. Journal of Fish Biology 39: 295–300

Mackay I and Mann KH (1969) Fecundity of two cyprinid fishes in the River Thames, reading, England. Journal of Fisheries Research Board of Canada 26: 2795–2805

Mann RHK (1978) Observations on the biology of the perch, *Perca fluviatilis*, in the River Stour, Dorset. Freshwater Biology 8: 229–239

Mann RHK, Mills CA and Crisp DT (1984) Geographical variation in the life-history tactics of some species of freshwater fish. In: Potts GW and Wootton RJ (eds) Fish Reproduction: Strategies and Tactics, pp 171–186. Academic Press, London

Mills CA and Eloranta A (1985) The biology of *Phoxinus phoxinus* (L.) and other littoral zone fishes in Lake Konnevsi, central Finland. Annales Zoologici Fennici 22: 1–12

Minckley WL and Meffe GK (1987) Differential selection by flooding in stream-fish communities of the arid American Southwest. In: Matthews WJ and Heins DC (eds) Community and Evolutionary Ecology of North American Stream Fishes, pp 93–104. University of Oklahoma Press, Norman

Miñano PA, Oliva-Paterna FJ, Fernández-Delgado C and Torralva M (2000) Edad y crecimiento de *Barbas graellsii* Steindachner, 1866 y *Chondrostoma miegii*, Steindachner, 1866 (Pisces, Cyprinidae) en el río Cinca (Cuenca Hidrográfica del Ebro, NE España). Miscellània Zoològica 23.2: 9–19

Moyle PB (1995) Conservation of native freshwater fishes in the Mediterranean-type climate of California, USA: a review. Biological Conservation 72: 271–279

Moyle PB, Smith JJ, Daniels RA and Baltz DM (1986) Distribution and ecology of the fishes of Sacramento-San Joaquín drainage system, California: a review. University of California Publications of Zoology 115: 225–256

Oberdorff T, Pont D, Hugueny B and Porchers JP (2002) Deveolpment and validation of a fish-based index for the assessment of 'river health' in France. Freshwater Biology 47: 1720–1734

Oliva-Paterna FJ, Torralva MM and Fernández-Delgado C (2002) Age, growth and reproduction of *Cobitis paludica* in a seasonal stream. Journal of Fish Biology 60: 389–404

Penaz M and Kokes J (1981) Notes on the diet, growth and reproduction of *Carassius auratus gibelio* in two localities in Southern Slovakia. Folia Zoologica 30: 83–94

Pimpicka E (1990) Formation of fecundity of tench *Tinca tinca* (L.) females in Lake Drweckie. Acta Ichthyologica et Piscatoria 20: 53–75

Rincón PA and Lobón-Cerviá J (1989) Reproductive and growth strategies of the red roach, *Rutilus arcasii* (Steindachner, 1866), in two contrasting tributaries of the River Duero, Spain. Journal of Fish Biology 34: 687–705

Ross ST (1991) Mechanisms structuring stream fish assemblages: are there lessons from introduced species? Environmental Biology of Fishes 30: 359–368

Sostoa A de (1990) Història natural dels països Catalans: peixos. [Natural history of Catalan countries: fishes.] Enciclopèdia Catalana, Barcelona, 487 pp

Vargas MJ and Sostoa A de (1997) Life-history pattern of the Iberian toothcarp *Aphanius iberus* (Pisces, Cyprinodontidae) from a Mediterranean estuary, the Ebro Delta (Spain). Netherlands Journal of Zoology 47: 143–160

Verneaux J (1981) Les poisons et la qualité des cours d'eau. Annales Scientifiques de l'Université de Franche-Comté 2: 33–41

Vila-Gispert A and Moreno-Amich R (2000) Fecundity and spawning mode of three introduced fish species in Lake Banyoles (Catalunya, Spain) in comparison with other localities. Aquatic Sciences 61: 154–166

Vila-Gispert A, Moreno-Amich R and García-Berthou E (2002) Gradients of life-history variation: an intercontinental comparison of fishes. Reviews in Fish Biology and Fisheries 12: 417–427

Viñolas D (1986) Biologia i ecologia de *Blennius fluviatilis* (Asso, 1801) en el riu Matarranya. PhD Thesis. Department of Zoology, University of Barcelona, 231 pp

Williamson MH and Fitter A (1996) The characters of successful invaders. Biological Conservation 78: 163–170

Winemiller KO (1989) Patterns of variation in life-history among South American fishes in periodic environments. Oecologia 81: 225–241

Winemiller KO and Rose KA (1992) Patterns of life-history diversification in North American fishes: implications for population regulation. Canadian Journal of Fisheries and Aquatic Sciences 49: 2196–2218

Wright RM and Schoesmith EA (1988) The reproductive success of pike, *Esox lucius*: aspects of fecundity, egg density and survival. Journal of Fish Biology 33: 623–636

Biological Invasions (2005) 7: 117–125

Genetic introgression on freshwater fish populations caused by restocking programmes

María José Madeira[1], Benjamín J. Gómez-Moliner[1,*] & Annie Machordom Barbé[2]
[1]*Facultad de Farmacia, Departamento de Zoología y Dinámica Celular Animal, Universidad del País Vasco, c/ Paseo de la Universidad, 7, 01006 Vitoria, Spain;* [2]*Museo Nacional de Ciencias Naturales (CSIC), Departamento de Biodiversidad y Biología Evolutiva, c/ José Gutiérrez Abascal, 2, 28006, Madrid, Spain;* **Author for correspondence (e-mail: ggpgomob@vc.ehu.es; fax: +34-945-013014)*

Received 4 June 2003; accepted in revised form 30 March 2004

Key words: brown trout, genetic introgression, LDH-C*, management, mtDNA, PCR-RFLP

Abstract

The brown trout (*Salmo trutta* L.) is one of the best studied native salmonids of Europe. Genetic studies on this species suggest that a large proportion of the evolutionary diversity corresponds to southern European countries, including the Iberian Peninsula, where this study is focused. Stocking activities employing non-indigenous hatchery specimens together with the destruction and fragmentation of natural habitats are major factors causing a decrease of native brown trout populations, mostly in the Mediterranean basins of the Iberian Peninsula. The main aim of the present work is to examine the genetic structure of the brown trout populations of the East Cantabrian region, studying the consequences of the restocking activities with foreign hatchery brown trout specimens into the wild trout populations. We have based our study on the Polymerase Chain Reaction and Restriction Fragment Length Polymorphism technique conducted on a mitochondrial fragment of 2700 base pairs and on the lactate dehydrogenase locus of the nuclear DNA. Our results show higher introgression rates in the Ebro (Mediterranean) basin than in the Cantabrian rivers.

Introduction

Most European countries have ratified the Convention on Biological Diversity (Rio de Janeiro 1992), which stresses that 'States are responsible for conserving their biological diversity and for using their biological resources in a sustainable manner.' Since then, biodiversity conservation and survival of species has became a priority for the people involved in the management of natural resources. This principle is more difficult to be applied to those species under exploitation mainly due to economic interests. In this context, a great number of freshwater fishes are seriously threatened with extinction as a direct consequence of human activities: overexploitation by fishing, habitat alteration, introduction of exotic fishes, or stocking activities (Ferguson 1990; Rhymer and Simberloff 1996; Elvira 1998; Cross 2000).

Stocking activities (introductions or reintroductions) with fishes grown in captivity has become a common practice in many countries with the primary aim of getting an increment of angling as well as for the rehabilitation of natural populations. It has been shown that restocking programmes can result in deleterious effects on the natural fish populations that in many cases are the same as those caused by introduction of exotic species: diseases and competition for food and habitat (Lynch and O'Hely 2001). All this might cause a displacement of the local populations, or in extreme cases, their extinction. Furthermore, introduction of fishes of the same species (stocking) but from a different geographical origin may introduce genetic changes in native populations due to the introgression of allochthonous genes on wild populations (Huxel 1999). It has been shown that natural hybridization has played an important role in the evolution of many animal taxa. However, stocking activities are increasing an anthropogenic hybridization which may reduce the genetic variability of the native populations and in consequence constrains the possibility for future adaptations (Allendorf et al. 2001). In addition, genetic introgression may cause a homogenization of wild populations from different geographical areas because, in most cases, the hatchery stocks form only a part of the total genetic variability of the species.

Brown trout (*Salmo trutta* L.) populations provide an interesting case to study the influence of stocking activities on native populations. Several studies using different genetic markers have demonstrated the existence of a considerable degree of genetic differentiation between brown trout populations throughout its natural distribution range, both at the macro and microgeographical levels (Ferguson 1989; Bernatchez et al. 1992; Hansen and Loeschcke 1996; Hynes et al. 1996). Important differences in the genetic trout diversity between Mediterranean and Atlantic drainages have also been reported (García-Marín et al. 1996). The locus LDH-C* is one of the most important diagnostic genetic marker systems to deal with the study of the population genetics of this species (Ferguson and Fleming 1983; Hamilton et al. 1989). Two major groups of brown trout have been described on the basis of the presence of LDH-C*90 and LDH-C*100 alleles: 'modern' (North Atlantic) and 'ancient' (South Atlantic and Mediterranean) lineages, respectively (Hamilton et al. 1989; García-Marín et al. 1999). Upon mtDNA analysis, Bernatchez et al. (1992) have described five major phylogenetic groups in Europe: Adriatic, Danube, Mediterranean, marmoratus and Atlantic. Furthermore, it has been shown that a large proportion of the evolutionary diversity of brown trout is sited in the south of Europe, including the Iberian Peninsula, where this study is focused (Giuffra et al. 1994; García-Marín and Pla 1996; García-Marín et al. 1999). Some authors have claimed that the Iberian Peninsula acted as a refuge during the last glaciations and proposed the existence of two distinct regional groups associated with Atlantic and Mediterranean drainages. Furthermore, Machordom et al. (2000) have demonstrated the presence of at least five distinct groups in the Iberian Peninsula trout populations: Mediterranean, Andalusian, Atlantic, Duero, and Cantabrian lineages.

Because of economic and sport-fishery interest, the Iberian rivers have been intensively restocked during the last decades employing hatchery strains of brown trout of exogenous origin, mainly from central and northern Europe. Although these stocking activities have been carried out for several decades, most studies indicate that introgression of native populations is limited (McNeil 1991). However, partial displacement of some native populations by hybridization and introgression is very common (Hindar et al. 1991; Martínez et al. 1993).

The aim of the present work was to evaluate the consequences of the restocking activities with allochthonous brown trout specimens into the wild populations. The study area was located in the north of the Iberian Peninsula and included two different drainages: the Mediterranean and Cantabrian–Atlantic basins.

This study was based on Polymerase Chain Reaction-Restriction Fragment Length Polymorphism (PCR-RFLP) analyses of the NADH dehydrogenase subunits 5/6 and the cytochrome b segment (ND 5/6-cyt b) of the mitochondrial genome, and the LDH-C* locus of the nuclear DNA. Most hatchery stocks are fixed or show a very high frequency of the allele LDH-C*90 which does not occur naturally in Iberian brown trout populations. In contrast, Iberian native populations are fixed to the LDH-C alleles *100 or *105. Thus, LDH-C* can be considered a good diagnostic marker to estimate the effects of stocking activities, such as genetic introgression, on wild population. On the other hand, the study of mtDNA haplotypes allows us to distinguish between brown trout populations of different geographic origins.

Materials and methods

Sampling

A total of 400 brown trouts were collected by electrofishing between 2001 and 2002 in 20 locali-

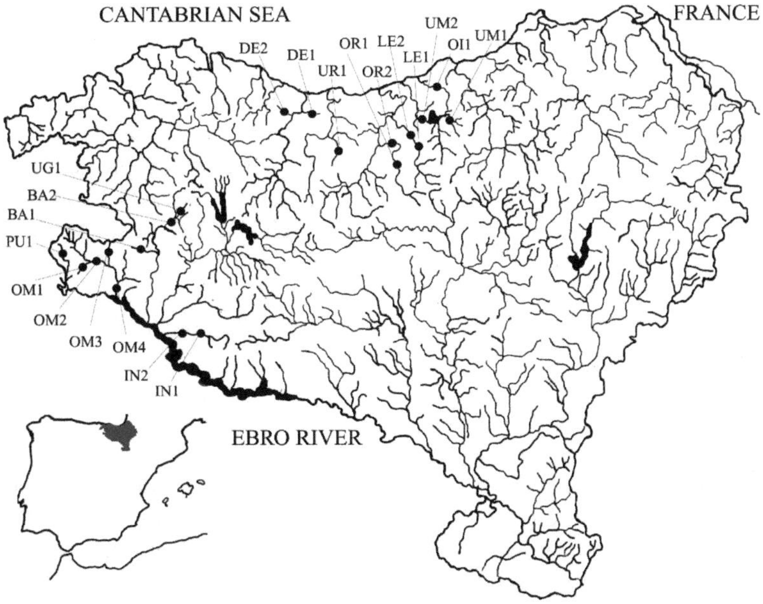

Figure 1. Codes and location of brown trout populations from Cantabrian and Mediterranean drainages in the Iberian Peninsula analysed in this study. See Table 1 for sampling characteristics. OI1: Oiartzun-Altzibar River; UM1: Urumea-Mendaraz River; UM2: Urumea-Urruzuno River; LE1: Leitzaran-Coto River; LE2: Leitzaran-Ameraun River; OR1: Oria-Araxes River (fishing preserve); OR2: Oria-Araxes River (fenced fishing area); UR1: Urola-Matxinbenta River; DE1: Deba-Sallabente River; DE2: Deba-Arranbide River.

ties from 11 different river basins from the north of the Iberian Peninsula (Figure 1). The sampling scheme included brown trout populations from river systems draining to the Cantabrian and Mediterranean (Ebro river system) basins. All of them have been restocked during decades with exogenous hatchery trout of central Europe origin. Two of the Mediterranean rivers included in this study have been regularly restocked up to the sampling years (OM2 and BA1). The rest of the populations have not been restocked since 1996. Samples from the two local fish farms (another 40 specimens), which have been used by local Governments for restocking activities have also been included.

Each captured fish was anaesthetized with etilenglieol (0.05%). A few scales were obtained and preserved in ethanol until their analysis. All fishes were released after handling, resulting in a non-lethal sampling procedure. Total genomic DNA was isolated from scales, using the standard phenol/chloroform–isoamylalcohol protocol after the proteolytic disruption of tissue with proteinase K. DNA was then quantified by fluorimetry and its purity estimated after migration on a 0.7% agarose gel.

Molecular analysis

The locus LDH-C* was PCR amplified according to the protocol reported by McMeel et al. (2001). The 440 bp amplified fragments were digested with the restriction endonuclease *Bsl* I and then electrophoresed on a 2% agarose gel using a 100 bp ladder as a molecular weight standard. According to McMeel et al. (2001), two alleles, LDH-C*90 and LDH-C*100, can be differentiated by this method.

mtDNA haplotypes were characterized by PCR amplification of a segment containing the subunits 5 and 6 of the NADH dehydrogenase gene (ND 5/6) and the cytochrome b gene. PCR conditions followed those described by Machordom et al. (2000). The 2700 bp amplicons were digested with five restriction enzymes (Machordom et al. 2000): *Alu* I, *Hinc* II, *Msp* I, *Rsa* I and *Sau* 3AI. Restriction fragments were visualized in a 2% agarose gel using a 100 bp ladder as a molecular weight standard, to calculate the molecular size of the restriction fragments. Each haplotype was defined by a five letter code following the RFLPs patterns described in Machordom et al. (2000).

According to previous studies carried out in the Iberian Peninsula, wild populations have the diagnostic allele LDH-C*100 fixed in their genetic composition. In this manner, the percentage of introgression was calculated from the mean percentage of the central European alleles at this locus (LDH-C*90).

Results

LDH-C analysis*

Three different genotypes have been observed from the analysis of the LDH-C* locus. LDH-C*90/90 is specific for hatchery stocked brown trout whereas LDH-C*100/100 characterizes Iberian native populations. The heterozygous LDH-C*100/90 genotype identifies hybrid specimens. The whole hatchery specimens of the two stocks studied showed LDH-C*90/90 genotype as was expected for the Central Europe origin of domestic foreign specimens. Heterozygous specimens were not detected in the hatchery stocks (Table 1).

The estimated introgression percentage was very variable among populations (0–65%). There were great differences between the mean introgressions of Cantabrian and Mediterranean populations (9.7 and 30.6%, respectively). The introgression percentage of those populations which were unstocked during the last 6 years was significantly higher in the Mediterranean than in the Cantabrian rivers (Table 1).

Only three of the analysed populations were free of introgression (OI1, UN1, and DE1). Another four showed residual levels of introgression by foreign alleles of hatchery origin (UM2, LE1, OR1 and DE2; LDH-C*90 frequency < 5%). All these populations were located in the Cantabrian region where the domestic foreign alleles occurred at low frequencies (LDH-C*90 frequency < 10%). However, the River Urumea exhibited the highest level of introgression for this region (22%). Hatchery diagnostic allele (LDH-C*90) was not observed in homozygosis in any Cantabrian population; therefore, the introgression detected in this region was only due to hybrid specimens (LDH-C*100/90).

On the other hand, the foreign LDH-C*90 allele was observed in all the analysed Mediterranean localities, including those where the stocking practices stopped before 1996. Unlike the Cantabrian region, high values of introgression were found in the Mediterranean populations unstocked since 1996 (PU = 45%), showing both non-native genotypes LDH-C*90/90 and LDH-C*100/90 in several localities (OM1, OM3, BA2, and UG1). In those Mediterranean populations which were restocked until 2002 (ZA, OM4, and BA1), the estimated introgression rate ranged between 20 and 65%.

mtDNA analysis

A total of four haplotypes were detected among the 20 natural populations analysed. Using capital letters referring to RFLP pattern produced by each endonuclease in the order *Alu* I, *Hinc* II, *Msp* I, *Rsa* I and *Sau* 3AI, the four haplotypes found were AAAAA, ABACA, BBCDA, and BBAAA. According to Machordom et al. (2000), the AAAAA is the haplotype of hatchery reared trouts with Central Europe origin. This haplotype is also shared by the Cantabrian native populations of brown trout. ABACA is the general Iberian Atlantic haplotype and can be considered a diagnostic haplotype for the populations of the Cantabrian slope. BBCDA in the study region of this work is considered a native haplotype for the Iberian–Mediterranean populations. BBAAA is an undescribed haplotype which has been found in four populations of the Mediterranean slope (IN1, IN2, OM3, and OM4).

Every specimen of the two hatchery stocks analysed showed the AAAAA haplotype, which characterizes Atlantic and central European foreign brown trout. ABACA and AAAAA were nearly the only haplotypes present in the Cantabrian populations, with the former showing higher frequencies. Both haplotypes coexisted in several populations (UM1, UM2, LE1, LE2, OR1, OR2, DE1, and DE2). On the other hand, the ABACA haplotype was fixed in one population (OI1), and the AAAAA haplotype was fixed in another one (UR1). Unexpectedly, in the Deba river (DE2) from the Cantabrian slope, the Mediterranean haplotype BBCDA was found in a proportion of 20%. The LDH-C* marker confirmed the Iberian native genotype for these specimens (LDH-C*100/100).

The populations sampled at the Mediterranean drainage were composed of a mixture of the four

Table 1. mtDNA haplotypes inferred from RFLP and allele frequency data detected in the 20 samples of brown trout for Cantabrian and Mediterranean drainages analysed in this study and the two hatchery farms. The sampling areas are indicated as in Figure 1.

Drainage	River	mtDNA haplotype	Locus LDH-C* genotype	Frequencies	Stocking*	*I*
Cantabrian	OI1	ABACA	*100/100	100% (*n* = 20)	No	0%
	UM1	AAAAA	*100/100	40% (*n* = 8)	No	0%
		ABACA	*100/100	60% (*n* = 12)		
	UM2	AAAAA	*100/100	44% (*n* = 8)	No	2.7% (<5%)
		AAAAA	*100/90	6% (*n* = 1)		
		ABACA	*100/100	50% (*n* = 9)		
	LE1	AAAAA	*100/100	25% (*n* = 5)	No	2.5% (<5%)
		ABACA	*100/100	70% (*n* = 14)		
		ABACA	*100/90	5% (*n* = 1)		
	LE2	AAAAA	*100/100	39% (*n* = 7)	No	5.5%
		ABACA	*100/100	50% (*n* = 9)		
		ABACA	*100/90	11% (*n* = 2)		
	OR1	AAAAA	*100/100	40% (*n* = 8)	No	5%
		ABACA	*100/100	50% (*n* = 10)		
		ABACA	*100/90	10% (*n* = 2)		
	OR2	AAAAA	*100/100	60% (*n* = 12)	No	7.5%
		AAAAA	*100/90	10% (*n* = 2)		
		ABACA	*100/100	25% (*n* = 5)		
		ABACA	*100/90	5% (*n* = 1)		
	UR1	AAAAA	*100/100	55% (*n* = 11)	No	22.5%
		AAAAA	*100/90	45% (*n* = 9)		
	DE1	AAAAA	*100/100	25% (*n* = 5)	No	0%
		ABACA	*100/100	75% (*n* = 15)		
	DE2	ABACA	*100/100	75% (*n* = 15)	No	2.5%
		ABACA	*100/90	5% (*n* = 1)		
		BBCDA	*100/100	20% (*n* = 4)		
Mediterranean	OM1	BBCDA	*100/100	25% (*n* = 5)	No	45%
		BBCDA	*100/90	55% (*n* = 11)		
		BBCDA	*90/90	10% (*n* = 2)		
		AAAAA	*100/90	5% (*n* = 1)		
		AAAAA	*90/90	5% (*n* = 1)		
	OM2	BBCDA	*100/100	53% (*n* = 9)	No	23.5%
		BBCDA	*100/90	47% (*n* = 8)		
	OM3	BBCDA	*100/100	10% (*n* = 2)	No	45%
		BBCDA	*100/90	50% (*n* = 10)		
		BBCDA	*90/90	5% (*n* = 1)		
		AAAAA	*100/100	5% (*n* = 1)		
		AAAAA	*100/90	15% (*n* = 3)		
		BBAAA	*100/90	15% (*n* = 3)		
	OM4	BBCDA	*100/100	10% (*n* = 1)	Yes	65%
		BBCDA	*100/90	30% (*n* = 3)		
		BBCDA	*90/90	20% (*n* = 2)		
		BBAAA	*100/100	10% (*n* = 1)		
		BBAAA	*90/90	30% (*n* = 3)		
	BA1	BBCDA	*100/100	66% (*n* = 10)	Yes	20%
		BBCDA	*100/90	20% (*n* = 3)		

Table 1. Continued.

Drainage	River	mtDNA haplotype	Locus LDH-C* genotype	Frequencies	Stocking*	*I*
		AAAAA	*100/90	7% (*n* = 1)		
		AAAAA	*90/90	7% (*n* = 1)		
	BA2	BBCDA	*100/100	65% (*n* = 13)	No	17.5%
		BBCDA	*100/90	20% (*n* = 4)		
		BBCDA	*90/90	5% (*n* = 1)		
		AAAAA	*100/100	5% (*n* = 1)		
		AAAAA	*100/90	5% (*n* = 1)		
	UG1	BBCDA	*100/100	39% (*n* = 6)	No	36.6%
		BBCDA	*100/90	7% (*n* = 1)		
		BBCDA	*90/90	20% (*n* = 3)		
		AAAAA	*100/100	7% (*n* = 1)		
		AAAAA	*100/90	27% (*n* = 4)		
	IN1	BBAAA	*100/100	49% (*n* = 4)	No	12.5%
		BBAAA	*100/90	13% (*n* = 1)		
		ABACA	*100/100	25% (*n* = 2)		
		ABACA	*100/90	13% (*n* = 1)		
	IN2	BBAAA	*100/100	42% (*n* = 8)	No	2.5%
		BBAAA	*100/90	5 (*n* = 1)		
		AAAAA	*100/100	5 (*n* = 1)		
		ABACA	*100/100	48% (*n* = 9)		
	PU1	BBCDA	*100/90	35% (*n* = 7)	No	45%
		AAAAA	*100/100	15% (*n* = 3)		
		AAAAA	*100/90	45% (*n* = 9)		
		AAAAA	*90/90	5% (*n* = 1)		
Hatchery strain	HAT1	AAAAA	*90/90	100% (*n* = 20)		
	HAT2	AAAAA	*90/90	100% (*n* = 20)		

I = introgression index related to the LDH-C*90 allele frequency.
*No: unstocked during the last 6 years; Yes: stocked until 2002.

identified mtDNA haplotypes: AAAAA, ABACA, BBCDA, and BBAAA. The stocked populations as well as many of those which were not stocked after 1996 (OM1, OM3, BA1, BA2, UG1, IL1, and PU1) showed different degrees of genetic contamination with the hatchery haplotype AAAAA. Moreover, the ABACA haplotype was found in the Inglares river (IN1, IN2). The LDH-C* analysis indicated the presence of LDH-C*100/100 genotype in most of these 'ABACA specimens'.

Discussion

The three genotypes obtained from the analysis of the LDH-C* locus in the populations studied allow us to differentiate native and stocked brown trouts. The genotype LDH-C*90/90 indicates stocked brown trouts because it was present in all the hatchery samples analysed. According to other studies (García-Marín et al. 1999), the Iberian native populations are homozygous to the allele LDH-C*100. In view of the absence of heterozygotes in the hatchery stocks analysed, we can considerer the heterozygous specimens as hybrids between native and stocked trouts.

Considerably higher introgression rates were detected in the Ebro river (Mediterranean) populations when compared with those of the Cantabrian region. Our results are in agreement with those of other published studies on genetic introgression between wild and stocked brown trouts in Spanish rivers. Thus, the low introgression observed in the Cantabrian rivers was similar and congruent with those found in other rivers located in the north (Morán et al. 1991, 1995;

García-Marín and Pla 1996). Similarly, studies carried out in other Ebro tributaries (García-Marín and Pla 1996; Machordom et al. 1999) have showed introgression and admixture rates equivalent to those found in this study.

These results suggested that environmental conditions of rivers might explain the observed differences in introgression and admixture between the study regions as previously suggested by Almodóvar et al. (2001). In this context, Cantabrian rivers present some common characteristics that would suppose a disadvantage for hatchery stocked trouts. They are short rivers belonging to small basins of narrow valleys with marked torrential character. The average gradient is 11%; they are divided into rapids, deeper sections, and pools. In contrast, Mediterranean rivers present an average slope of 2%; they are seasonal rivers with a low water level and warm temperatures in summer. These environmental differences among rivers of both regions seem to have played an important role and could explain the introgression rate variation detected. It has been shown that the type of environment in which hatchery trout is stocked is important for their survival and reproduction (Poteaux et al. 1998). The variable success of survival and reproduction of stocked fishes is probably the main factor explaining the different introgression rates observed between both regions and could also explain the indirect effect of restocking on the structure of the populations (Poteaux et al. 1998).

On the other hand, several studies reported poor survival and reproduction of stocked trouts in those rivers where high levels of natural reproduction of native trout population occurred (Morán et al. 1991). In this way, trout fry number as well as whole trout populations were much higher in the Cantabrian basin than in the Mediterranean streams analysed. This indicates a higher level of natural reproduction in Cantabrian rivers which could have a negative effect on the survival percentage of stocked trouts.

Although the Cantabrian rivers were strongly restocked over the past decades, this region presented a low introgression rate. Therefore, cessation of stocking for six or more years seemed to have caused a progressive reduction of domestic foreign allele frequencies in the wild populations. However, in the Mediterranean rivers where restocking activities were stopped for the last 6 years, the introgression index was still high. The highest introgression percentages (65%) corresponded to those populations restocked until 2002, probably due to the earliest introductions. These populations might constitute a reservoir of allochthonous genes of hatchery origin that could migrate to other river sections or streams, and cause a negative effect on the genetic integrity of the wild populations.

The observed pattern of haplotype distribution demonstrated great differentiation between the populations of the Cantabrian and Mediterranean regions. The complementary use of mtDNA provided useful additional information about genetic introgression to that obtained with the LDH-C* locus.

The introgression index observed in some Mediterranean populations might have been underestimated by assuming that all sampled trouts with LDH-C*100/100 genotype were native. Some of these 'native trout' presented the AAAAA haplotype and were homozygous for the autochthonous diagnostic allele LDH-C*100. If it is assumed that this haplotype (AAAAA) is non-native for Mediterranean populations, these specimens have to be considered hybrids between hatchery and native breeders. This admixture between a native nuclear genotype and a non-native mitochondrial haplotype might be explained by breeding among heterozygous (LDH-C*100/90) or homozygous (LDH-C*100/100) female hybrids (F1 or F2 proceeding from a hatchery female trout, respectively), and males (both with homo- or heterozygous genotype). Those specimens showing native mitochondrial haplotype (BBCDA or BBAAA) together with allochthonous nuclear allele (LDH-C*90) could have the same hybrid origin. All these results confirmed a high interbreeding index between wild and stocked trouts and their descendants in this region.

The presence of the Mediterranean BBCDA haplotype in one river of the Cantabrian region is difficult to explain. It seems to be a genetic contamination, but its hatchery origin must be dismissed because this haplotype was not detected in the hatchery stocks used for local restocking. It might have come from introductions carried out by uncontrolled anglers from hatcheries working with native Mediterranean trouts, or translocated directly from a Mediterranean river. Another explanation is that the presence of this Mediterranean haplotype could have been produced by a natural river capture. These natural captures have been demonstrated in different natural systems, and also in the Cantabrian region (Alonso-Otero 1986). In the area studied, Cantabrian and Mediterranean headwaters are close together, and this could facilitate this natural process.

On the opposite side, the ABACA haplotype has been considered until now a Cantabrian and Atlantic hapotype (Machordom et al. 2000). Its presence in the Ebro tributaries cannot be seen as a result of stocking activities because it was not detected in the hatchery stocks analysed. The high proportion estimated in the Inglares river (Mediterranean region) suggested that this haplotype could occur naturally in the native brown trout populations of some upper tributaries of the Ebro basin. Furthermore, a new mtDNA haplotype (BBAAA) was also found in this river and in other close Ebro tributary rivers. These two haplotypes were not detected in other lower tributaries of the Ebro basin. Therefore, their presence appears to be restricted to the upper tributaries of this basin and could indicate the presence of an Iberian-Mediterranean refuge in the Ebro basin. On the other hand, the presence of particular haplotypes in this region may also be due to other reasons as a genetic local adaptation to the particular environment conditions (Taylor 1991) of these rivers. As previously stated by Suarez et al. (2001), the Iberian brown trout maintains a genetic diversity different from that shown by other European trout populations. Restocking policies should take into account several guidelines (Almodóvar et al. 2001) for conservation and management of the Evolutionary Significant Units (ESUs) of native brown trout populations in the Iberian Peninsula rivers.

In conclusion, from the management and conservation points of view, it is obvious that any kind of restocking with foreign specimens should always be avoided to preserve the autochthonous genetic pools. Despite stocking activities being stopped during the last 6 years, the presence of domestic foreign allele (LDH-C*90) demonstrates that hatchery brown trouts reproduced under natural conditions. On the other hand, the Cantabrian rivers' natural conditions seem to prevent or eliminate the negative consequences of such introductions more accurately. Thus, the recovery of the native genetic characteristics of these Cantabrian populations could be possible if the management was adequate.

Acknowledgements

This work was supported financially by the 'Diputación de Álava', 'Diputación de Gipuzkoa' and the 'Departamento de Educación Universidades e Investigación' of the Basque Country which provided a grant to M.J. Madeira. We are grateful to D. Ramiro Asensio from the 'Federación de Pesca de Álava', for his valuable technical assistance in the electrofishing activities. Samples from Gipuzkoa rivers have been obtained by EKOLUR and the local technical staff.

References

Allendorf F, Leary R, Spruell P and Wenburg J (2001) The problems with hybrids: setting conservation guidelines. Trends in Ecology and Evolution 16 (11): 613–619

Almodóvar A, Suarez J, Nicola GG and Nuevo M (2001) Genetic introgression between wild and stocked brown trout in the Douro River basin, Spain. Journal of Fish Biology 59(Supplement A): 68–74

Alonso-Otero F (1986) La erosión fluvial en la divisoria cantábrica. El problema de las capturas: la cuenca del Urola. In: Alianza Editorial (ed) Atlas de geomorfología, pp 178. Madrid, Spain

Bernatchez L, Guyomard R and Bonhomme F (1992) DNA sequence variation of the mitochondrial control region among geographically morphologically remote European brown trout *Salmo trutta* populations. Molecular Ecology 1: 161–173

Cross T (2000) Genetic implications of translocation and stocking of fish species, with particular reference to western Australia. Aquaculture Research 31(1): 83–94

Elvira B (1998) Impact of introduced fish on the native freshwater fish fauna of Spain. In: Cowx IG (ed) Stocking and Introduction of Fish, pp 186–190. Fishing News Books, Oxford

Ferguson A (1989) Genetic differences among brown trout, *Salmo trutta*, stocks and their importance for the conservation and management of the species. Freshwater Biology 21: 35–46

Ferguson A and Fleming CC (1983) Evolutionary and taxonomic significance of protein variation in brown trout (*Salmo trutta fario*) and other salmonid fishes. In: Oxford G and Rollinson D (eds) Protein Polymorphism: Adaptative and Taxonomic Significance, pp 85–99. Academic Press, London

Ferguson MM (1990) The genetic impact of introduced fishes on native species. Canadian Journal of Zoology 68: 1053–1057

García-Marín JL and Pla C (1996) Origins and relationship of native populations of *Salmo trutta* (brown trout) in Spain. Heredity 77: 313–323

García-Marín JL, Utter FM and Pla C (1999) Postglacial colonization of brown trout in Europe based on distribution of allozyme variants. Heredity 82: 46–56

Giuffra E Bernatchez L and Guyomard R (1994) Mitochondrial control region and protein coding genes sequence variation among phenotypic forms of brown trout *Salmo trutta* from northern Italy. Molecular Ecology 3: 161–171

Hamilton KE, Ferguson A, Taggart JB, Tomasson T, Alker A and Fahy E (1989) Post-glacial colonization of brown trout, *Salmo trutta* L. Ldh-5 as a phylogeographical marker locus. Journal of Fish Biology 35: 651–664

Hansen MM and Loeschcke V (1996) Genetic differentiation among Danish brown trout populations, as detected by RFLP analysis of PCR amplified mitochondrial DNA segments. Journal of fish Biology 48: 422–436

Hindar K, Ryman N and Utter F (1991) Genetic effects of cultured fish on natural fish populations. Canadian Journal of Fisheries and Aquatic Sciencies 48: 945–957

Huxel G (1999) Rapid displacement of native species by invasive species: effects of hybridization. Biological Conservation 89: 143–152

Hynes RA, Ferguson A and McCann MA (1996) Variation in mitochondrial DNA and post-glacial colonization of north western Europe by brown trout. Journal of Fish Biology 48: 54–67

Lynch M and O'Hely M(2001) Captive breeding and the genetic fitness of natural populations. Conservation Genetics 2: 363–378

Machordom A, García-Marín JL, Sanz N, Almodóvar A and Pla C (1999) Allozyme diversity in brown trout (*Salmo trutta*) from Central Spain: genetic consequences of restocking. Freshwater Biology 41: 707–717

Machordom A, Suárez J, Almodóvar A and Bautista JM (2000) Genetic differentiation and phylogenetic relationships among Spanish brown trout (*Salmo trutta*) populations. Molecular Ecology 9: 1325–1338

Martínez P, Arias J, Castro J and Sánchez L (1993) Differential stocking incidence in brown trout (*Salmo trutta*) populations from Northwestern Spain. Aquaculture 114: 203–216

McMeel O, Hoey E and Ferguson A (2001) Partial nucleotide sequences, and routine typing by polymerase chain reaction-restriction fragment length polymorphism, of the brown trout (*Salmo trutta*) lactate dehydrogenase, LDH-C1*90 and *100 alleles. Molecular Ecology 10: 29–34

McNeil WJ (1991) Expansion of cultured Pacific salmon into marine ecosystems. Aquaculture 98: 173–183

Morán P, Pendás AM, García-Vázquez E and Izquierdo JJ (1991) Failure of stocking policy, of hatchery reared brown trout, *Salmo trutta* L., in Asturias, Spain, detected using LDH-5*as a genetic marker. Journal of Fish Biology 39: 117–122

Morán P, Pendás AM, García-Vázquez E, Izquierdo JJ and Lobón-Cerviá J (1995) Estimates of gene flow among neighbouring populations of brown trout. Journal of Fish Biology 46: 593–602

Poteaux C, Bonhomme F and Berrebi P (1998) Differences between nuclear and mitochondrial introgression of brown trout populations from a restocked main river and its unrestocked tributary. Biological Journal of the Linnean Society 63: 379–392

Rhymer JM and Simberloff D (1996) Extinction by hybridisation and introgression. Annual Review of Ecology and Systematics 27: 83–109

Río de Janeiro (1992) Text of the Convention on Biological Diversity. Section I, Article 2, 5–6. Secretariat of the Convention on Biological Diversity. United Nations Environment Programme

Suarez J, Bautista JM, Almodóvar A and Machordom A (2001) Evolution of the mitochondrial control region in Palaearctic brown trout (*Salmo trutta*) populations: the biogeographical role of the Iberian Peninsula. Heredity 87: 198–206

Taylor EB (1991) A review of local adaptation in Salmonidae, with particular reference to Pacific and Atlantic salmon. Aquaculture 98: 185–207

Biological Invasions (2005) 7: 127–133

Eradications of invasive alien species in Europe: a review

Piero Genovesi
Chair European Section IUCN SSC Invasive Species Specialist Group, National Wildlife Institute, Via Ca' Fornacetta 9, 40064 Ozzano, Emilia (BO), Italy (e-mail: infspapk@iperbole.bologna.it; fax: +39-051-796628)

Recevied 4 June 2003; accepted in revised form 30 March 2004

Key words: animal rights, biological invasions, control, mitigation of impacts, removal

Abstract

Eradication of alien species is a key conservation tool to mitigate the impacts caused by biologic invasions. The aim of the present paper is to review the eradications successfully completed in Europe and to discuss the main limits to a wider application of this management option in the region. On the basis of the available literature – including conference proceedings, national reports to the Bern Convention, etc. – a total of 37 eradication programmes have been recorded. Thirty-three eradications were carried out on islands and four on the mainland. The rat (*Rattus* spp.) has been the most common target ($n = 25$, 67%), followed by the rabbit ($n = 4$). In many cases, these eradications determined a significant recovery of native biodiversity. Differently to other regions of the world, no eradications of alien invertebrates and marine organisms have been recorded; regarding invasive alien plants, it appears that only some very localized removals have been completed so far in Europe. The limited number of eradications carried out in Europe so far is probably due to the limited awareness of the public and the decision makers, the inadequacy of the legal framework, and the scarcity of resources. Synthetic guidelines for improving the ability of European states to respond to aliens incursions are presented.

Introduction

Eradication of alien species is globally acknowledged as a key management option for mitigating the impacts caused by biological invasions (e.g., Wittenberg and Cock 2001; Genovesi and Shine 2003). The Convention on Biological Diversity (CBD) calls for a hierarchical approach, primarily based on the prevention of unwanted introductions, but considering eradication as the best alternative when prevention fails (guiding principles adopted in 2002 with Decision VI/23). Article 11 of the Bern Convention (which has almost all European states as its members) calls parties to strictly control the introduction of invasive alien species, and the standing committee of this convention has approved many recommendations urging parties to activate eradications of introduced species. Also, the Global Strategy for Plant Conservation, adopted by the CBD Conference of the Parties in 2002, urges parties to eradicate some major alien species that threaten plants, plant communities and associated habitats and ecosystems.

Many invasive alien species have been eradicated worldwide, managing in this way to prevent the impacts they cause to biological diversity, economy and human well being (Simberloff 2002). In recent years, eradications have become a routine management tool especially on islands, where many introduced vertebrates have been successfully removed: for example, in New Zealand, 156 eradications have been completed (D. Veitch, pers. comm.); in northwestern Mexico, eradications have been carried out from 23 islands since 1995 (Tershy et al.

2002); in West Australia, mammal eradications have been completed on 48 islands since 1969 (Burbridge and Morris 2002). Most of these eradications have involved vertebrates, but there are also examples of successful eradications of plants (Rejmanek and Pitcairn 2002) and invertebrates, including fruit flies from Nauru (Allwood et al. 2002) and *Anopheles gambiae* from over 30,000 kmq of Brazil in the 1950s (Davis and Garcia 1989). Even marine organisms have been eradicated in some cases (when invasion was still localized) as a Mussel (*Mytilopsis* sp.) introduced in Cullen Bay (Australia) (Bax et al. 2002) and a sabellid polychaete (*Terebrasabella heterouncinata*) successfully removed from a mariculture facility in California (Galil 2002).

Successful eradications have brought very significant effects in terms of recovery of native biological diversity. Focusing on Europe, the eradication of rats from the islands of the Mediterranean has been proven to determine the recovery of many colonial nesting seabirds as the storm petrel (*Hydrobates pelagicus*) and the Cory's shearwater (*Calonectris diomedea*) (Martín et al. 2000), but also of several terrestrial bird species such as the dunnock (*Prunella modularis*), the wren (*Troglodytes troglodytes*) and the rock pipit (*Anthus petrosus*) (Kerbiriou et al. 2004). In an analysis of the consequences of the black rat (*Rattus rattus*) introduction in the Mediterranean, Martin et al. (2000) concluded that the elimination of rats from medium-sized Mediterranean islands may be particularly efficient in recovering several bird species, some of which are highly threatened.

Surprisingly, despite removal techniques having been greatly improved, Europe has a particularly solid technical and scientific background, and there are large areas (e.g., Mediterranean islands, Macaronesian archipelagos, etc.) where this management option may be very helpful in the recovery of threatened species and ecosystems – the eradication of alien species appears to be still only occasionally considered for conservation in Europe. For example, in the proceedings of a recent international conference on eradication of island invasives, there was only one European contribution reporting a failed attempt to eradicate *Spartina anglica* from some estuarine areas of Northern Ireland (Hammond and Cooper 2002).

In order to assess the diffusion of eradication programmes in Europe and the main limiting factors to a wider application of this management option, the present paper reviews all the known cases of eradications of alien species of plants and animals successfully completed in the region.

Methods

This review intends to cover all cases of eradications, defined as the complete and permanent removal of all wild populations of a species from a defined area by means of a time-limited campaign (Genovesi 2000; modified from Bomford and O'Brien 1995). The above definition is considered in an extended way, including the cases of removal of few individuals, because I did not want to exclude the cases when an introduced species was detected early after its arrival and immediately removed. This, in fact, is the best response to an introduction, although such a large definition may risk to create some overlap between the eradication of a few individuals (e.g., removal of a few beavers in France or of a few individuals of domestic cats from an island, both cases included in the review) and the simple capture of individuals after escape from captivity (not included in the review).

The information on eradications is particularly difficult to collect, because it is scattered, often available only in grey literature, if it is published at all (Simberloff 2002). For the present review, I firstly checked the scientific publications, focusing in particular on the proceedings of conferences and workshops held in recent years on the issue of alien species (e.g., in 1999, the Council of Europe organized in Malta the first European workshop specifically focused on the eradication of terrestrial alien species). Another important source of information has been the reports prepared by parties of the Council of Europe to the Bern Convention secretariat. In fact, as said above, the Bern Convention Standing Committee has approved several recommendations (rec. 18 (1989); rec. 45 (1995); rec. 57 (1997); rec. 61 (1997); rec. 78 (1999); rec. 77 (1999); etc) urging parties to eradicate alien species threatening native biodiversity; to respond to these recommendations, many European countries have established national reports illustrating the activities carried out in their territories on this specific management option. In order to facilitate the collection of these national reports, the Council of Europe has also organized several meetings of experts in invasive alien species, and the proceedings of these workshops have also been extensively analysed for the review. Lastly, much information has been obtained directly by specialists, managers and NGOs.

Results

A total of 37 eradication programs successfully completed in Europe have been recorded (Table 1). These include the successful removal of the coypu and the muskrat from Great Britain, of rats, goats, rabbits and American minks from several small islands of the Macaronesia, Mediterranean, Brittany, Britain and the Baltic sea. Thirty-three eradications were carried out on islands and four on the mainland (muskrat, coypu and a small population of Indian porcupines from the British isles; the Canadian beaver from France). The rat (*Rattus* spp.) has been the most common target ($n = 25$, 67%), followed by the rabbit ($n = 4$). Almost all eradications realized in Europe (apart from the muskrat and the coypu) were started for conservation purposes; in 12 cases, the programmes were co-funded by the European Union through LIFE programmes. Most eradications were carried out after the 1980s ($n = 31$; 84%), and in recent years, the number of projects is rapidly increasing.

In some cases, the eradications reported have involved very few animals (e.g., cats from Alegranza, where only two animals were killed; 12 porcupines from Devon); very likely there have been many more cases than those reported in which a few individuals that arrived at some areas were rapidly removed.

I did not record any eradication of alien invertebrates and marine organisms. Although some local eradications of invasive plants have been carried out (for example in Great Britain by Plantlife International), I did not find information on these cases on the sources I acceded for the review. Apart from a few local cases (i.e. virtual eradication of Australian swamp stonecrop *Crassula helmsii* from a pond in Gerrard's Cross – Buckinghamshire – by using a glyphosate-based compound over 2 years; A. Miller, unpubl. report), I think it can be said that no larger eradication of alien plants has been ever successfully carried out in Europe so far.

Discussion

The list of eradications presented here is far from being comprehensive, as I probably failed to collect information on several small scale removals of alien plants and animals. However, the general picture that comes out from this review is probably correct. In Europe, only a very limited number of eradications have been successfully completed so far, and these do not include any invertebrate, plant or marine organisms. Europe, despite its long tradition of nature conservation, its solid scientific background and the large availability of funds in respect to other geographical regions, is in this specific field of action far behind other and less developed areas of the world.

The small number of eradications carried out in Europe is due to several reasons. For example, in the attempted eradication of the grey squirrel in Italy, the failure of the programme was mainly due to the inadequate legal basis (in most European states, alien species are often automatically protected by national laws), the scarce awareness, the unclear line of authority, and the opposition of radical animal rights groups (Genovesi and Bertolino 2001; Bertolino and Genovesi 2003). In the famous case of the *Caulerpa taxifolia* (Meinesz 1999), a decision on whether to start an eradication or not was delayed for long, partly because of an academic controversy, partly because of the unclear repartition of roles. The removal of the rabbit from a small island of the Canary was suspended when it was almost completed, because the project ran out of funds; a re-start of the eradication has now been approved, but the suspension has totally defeated the results obtained in the first campaign, when the population was almost completely removed (Martín 2002; A. Martín, pers. comm.). Legal inadequacy and scarce resources are indicated as the major constraints to the possibility to eradicate *Spartina anglica* from Northern Ireland estuaries (Hammond and Cooper 2002).

Table 1. Sources: (1) Gosling and Baker 1989; (2) J. Hughes, pers. comm.; (3) Zonfrillo 2002; (4) Smallshire and Davey 1989; (5) Macdonald et al. 2002; (6) Pascal 1999; (7) Kerbiriou et al. 2004; (8) M. Pascal, pers. comm.; (9) Rouland 1985; (10) Perfetti and Sposimo 2001; (11) Perfetti et al. 2001; (12) Oliveira 1999; (13) Pitta Groz 2002; (14) Jimenez 1994; (15) Martín 2002; (16) A. Martín, pers. comm.; (17) J.L. Rodriguez-Luengo, pers. comm.; (18) J. Mayol, pers. comm.; (19) Orueta, in preparation.

Country	Region	Archipelago	Island	Area (ha)	Species	Estimated population size	Eradication year	Source
Great Britain	Pertshire, Sussex			~100,000	*Ondatra zybethicus*		1935	1
	Wales		Puffin	32	*Rattus norvegicus*		1998	2
	Scotland		Handa	363	*Rattus norvegicus*		1997	2
	Scotland		Ailsa Craig	104	*Rattus norvegicus*		1990	2, 3
	Wales		Ramsey	253	*Rattus norvegicus*		2000	2
	West Anglia				*Myocastor coypus*	6000	1981	1
	Devon				*Hystrix brachyura*	12	1980	4
Estonia	Baltic		Hiiumaa	~100,000	*Mustela vison*	50	1998	5
France	Brittany	Sept Ile	Rouzic	3	*Rattus norvegicus*		1951	6
	Brittany	Sept Ile	Bono	22	*Rattus norvegicus*	700	1994	6
	Brittany	Sept Ile	aux Moines	9	*Rattus norvegicus*	200	1994	6
	Brittany	Sept Ile	Plate	5	*Rattus norvegicus*	100	1994	6
	Brittany	Sept Ile	aux Rats	0	*Rattus norvegicus*	20	1994	6
	Brittany	Rimains	Rimains	2	*Rattus norvegicus*	100	1994	6
	Brittany	Rimains	Chatellier	1	*Rattus norvegicus*	50	1994	6
	Brittany	Rimains	Rocher de Cancale	0	*Rattus norvegicus*	10	1994	6
	Brittany	Molène	Trielen	15	*Rattus norvegicus*	150	1996	6, 7
	Brittany	Molène	Enez ar C'hrizienn	1	*Rattus norvegicus*	30	1996	6
	Brittany	Houat	aux Chevaux	3	*Rattus norvegicus*		2002	8
	Brittany	Tomé	Tomé	30	*Rattus norvegicus*		2002	8
	Corsica		Lavezzi	110	*Rattus rattus*		2000	8
	St. Fargeau				*Castor canadensis*	24	1985	9
Italy	Tuscany	Tuscan	Legemini (1)	10	*Rattus rattus*		1999	10, 11
	Tuscany	Tuscan	Legemini (2)	10	*Rattus rattus*		2000	10, 11
	Tuscany	Tuscan	Scoglio La Peraiola	10	*Rattus rattus*		2000	10, 11
	Tuscany	Tuscan	dei Topi	10	*Rattus rattus*		2000	10, 11
	Tuscany	Tuscan	d'Ercole	10	*Rattus rattus*		2000	10, 11
	Tuscany	Tuscan	La Scola	2	*Rattus rattus*		2001	10, 11
Portugal	Macaronesia	Madeira	Deserta grande	1421	*Oryctolagus cuniculus*		1998	12
	Macaronesia	Azores	Praia Islet	11	*Oryctolagus cuniculus*	100–200	1997	13
Spain		Columbretes	Isla Grossa	14	*Oryctolagus cuniculus*	175	1993	14
	Macaronesia	Canary	Montana Clara	130	*Oryctolagus cuniculus*	127	2001	15
	Macaronesia		Lobos	430	*Felis catus*	4	2001	16, 17
	Macaronesia		Alegranza	1020	*Felis catus*	2	2002	16, 17
	Mediterranean	Balearic	Dragonera	280	*Capra hircus*		1975	18
	Mediterranean		Conills	1	*Rattus rattus*	>100	1999	18
	Mediterranean	Chafarines	Ray Francisco	12	*Rattus rattus*	ca. 50/ha	2000	19

The lack of concern, awareness and public support to the removal of vertebrates seems more diffuse in Europe than in other regions of the world. Apart from the grey squirrel case in Italy, several goat eradication projects have been stopped by public opposition (e.g., in the Parco Naturale di Portofino, Italy); public opposition is likely the main reason why only one goat eradication has been completed so far. Even in the case of the Coypu, the removal of a population recently introduced in a small lake in Sicily was strongly opposed by the local branch of the

WWF, and never started. Also, a proposal to eradicate the hedgehog (*Erinaceus europeus*) from Uist (Western Islands, Scotland), where the species causes impact on several bird species by egg predation, has been rejected for the ethical opposition to the control techniques.

Another problem is the limited ability to detect new invasions early and to rapidly respond to these. Although large scale eradications are scientifically and technically challenging, the best cases of eradication are those carried out rapidly after the arrival of a new species, before it starts to spread (e.g., Rouland 1985). In Italy, we recently discovered a population of Asian squirrels (*Callosciurus* sp.) in Maratea, a small tourist town on the southern coast of Italy; the squirrels have probably been introduced over 30 years ago, and in this time lapse, no local or national service (forest service, game departments, etc.) managed to detect the species; only in late 2002, when the population became very abundant causing increasing problems to trees and cables, the presence was reported to the National Wildlife Institute, when it was probably too late to remove them (G. Aloise et al., in preparation).

Despite the various problems highlighted here, there are several examples of very effective actions carried out in Europe. The eradication of the coypu from West Anglia is one of the largest and the most complex eradications ever realized in the world; its success was made possible by a science-based planning of the removal, adequate funding and the approval of a specific legislation. The eradication of the Himalayan porcupine from Devon required ca. €230,000 for removing 12 animals only (costs actualized to year 2000), but likely prevented much more severe economic losses to crops in the long term (Smallshire and Davey 1989). The ongoing eradication of the ruddy duck (*Oxyura jamaicensis*) from the Palearctic, is indeed the most ambitious eradication ever planned for conservation purposes, as it requires a complex coordination and cooperation scheme among many different countries, where the main control efforts need to be concentrated in a country (Great Britain), that is not the area where the impacts are recorded (hybridization with the white headed duck, *O. leucocephala*, occurs in the Iberian peninsula) (Hughes et al. 1999); furthermore, the control of the ruddy duck (a beautiful ornamental duck) in Great Britain shows that it is possible to effectively address the opposition of the public, provided a solid effort and commitment by both the academic world and the non-profit organisations are made.

In conclusion, on the basis of the information on eradication summarized in the present paper, the main lesson that we can learn is that in Europe, more than elsewhere, we urgently need to revise our policies to ensure early detection and rapid response to new incursions, with an increased ability to eradicate at least the most threatening alien species. The key elements for such a revision of national policies have been recently reviewed by Genovesi and Shine (2003) and include the following:

- Promote education and public awareness programmes to engage local communities and appropriate sector groups in eradication; encourage their participation.
- Review national legislation to ensure that the legal status of alien species is compatible with mitigation measures.
- Streamline the authorization process for rapid response; consider the use of emergency orders where urgent eradication action is needed; equip competent authorities with powers to take appropriate mitigation measures.
- Establish procedures to collect, analyse and circulate information of alien species, including identification keys for different taxonomic groups.
- Set up early warning systems, focusing especially on key areas.
- Prepare contingency plans for eradicating specific taxa (e.g., plants, invertebrates, marine organisms, fresh-water organisms, fresh-water fishes, reptiles, amphibians, birds, small mammals, large mammals).
- Provide adequate funds and equipment for rapid response to new invasions and train relevant staff to use the eradication methods selected.
- Prepare and implement, providing adequate funds and support, eradication plans for some major alien species.

Acknowledgements

The information in this paper was largely provided by many colleagues and institutions. I wish to thank Nicola Baccetti, Simon Baker, Giuseppe Brundu, Lois Child, Bruno Foggi, Martin

Harper (Plantlife International), Julian Hughes (RSPB), Juan Luis Rodriguez Luengo, Tiit Maran, Aurelio Martín, Joan Mayol, Amanda Miller, Paulo Jorge Oliveira, Jorge Orueta, Michael Pascal 'Ratator', Antonio Perfetti, Maria 'super' Pitta Groz. Special thanks to Bé Queiroz for having helped me to understand the elements of the European conservation policy.

References

Allwood AJ, Vueti ET, Leblanc L and Bull R (2002) Eradication of introduced *Batrocera* (Diptera: Tephritidae) in Nauru using male annihilation and protein bait application techniques. In: Veitch D and Clout M (eds) Turning the Tide: The Eradication of Invasive Species, pp 19–25. IUCN SSC Invasive Species Specialist Group. IUCN, Gland, Switzerland/Cambridge, UK, viii + 414 pp

Bax N, Hayes K, Marshall A, Parry D and Thresher R (2002) Man-made marinas as sheltered islands for alien marine organisms: establishment and eradication of an alien invasive marine species. In: Veitch D and Clout M (eds) Turning the Tide: the Eradication of Invasive Species, pp 26–39. IUCN SSC Invasive Species Specialist Group. IUCN, Gland, Switzerland/Cambridge, UK, viii + 414 pp

Bertolino S and Genovesi P (2003) Spread and attempted eradication of the grey squirrel (*Sciurus carolinensis*) in Italy, and consequences for the red squirrel (*Sciurus vulgaris*) in Eurasia. Biological Conservation 109: 351–358

Bomford M and O'Brien P (1995) Eradication or control for vertebrate pests? Wildlife Society Bulletin 23: 249–255

Burbridge AA and Morris KD (2002) Introduced mammal eradications for nature conservation on Western Australian islands: a review. In: Veitch D and Clout M (eds) Turning the Tide: the Eradication of Invasive Species, pp 64–70. IUCN SSC Invasive Species Specialist Group. IUCN, Gland, Switzerland/Cambridge, UK, viii + 414 pp

Davis JR and Garcia R (1989) Malaria mosquito in Brazil. In: Dahlsten DL and Garcia JR (eds) Eradication of Exotic Pests, pp 274–283. Yale University press, New Haven,-Connecticut

Galil BS (2002) Between serendipity and futility: control and eradication of aquatic invaders. Ballast Water News 11: 10–12

Genovesi P (2000) Guidelines for Eradication of Terrestrial Vertebrates: a European Contribution to the Invasive Alien Species Issue. Council of Europe, Strasbourg, tpvs65e-2000, 61 pp

Genovesi P and Bertolino S (2001) Human dimension aspects in invasive alien species issues: the case of the failure of the grey squirrel eradication project in Italy. In: McNeely JA (ed) The Great Reshuffling: Human Dimensions of Invasive Alien Species, pp 113–119. IUCN, Gland, Switzerland/ Cambridge, UK, vi + 242 pp

Genovesi P and Shine C (2003) European Strategy on Invasive Alien Species. Council of Europe, Strasbourg, t-pvs(2003)7 rev, 50 pp

Gosling LM and Baker SJ (1989) The eradication of muskrats and coypus from Britain. Biological Journal of the Linnean Society 38: 39–51

Hammond MER and Cooper A (2002) *Spartina anglica* eradication and inter-tidal recovery in Northern Ireland estuaries. In: Veitch D and Clout M (eds) Turning the Tide: the Eradication of Invasive Species, pp 124–131. IUCN SSC Invasive Species Specialist Group. IUCN, Gland, Switzerland/Cambridge, UK, viii + 414 pp

Hughes B, Criado J, Delany S, Gallo-Orsi U, Green AJ, Grussu M, Perennou C and Torrel JA (1999) The status of the Ruddy duck (*Oxyura jamaicensis*) in the western Paleartic and an action plan for eradication, 1999–2002. Report by the Wildfowl and Wetlands Trust to the Council of Europe

Jimenez J (1994) Gestione della fauna nelle piccole isole. In: Monbailliu X and Torne A (eds) La gestione degli ambienti costieri e insulari del Mediterraneo, pp 245–274. Medmaravis

Kerbiriou C, Pascal M, Le viol I and Garoche J (2004) Conséquences sur l'avifaune terrestre de lîle de Trielen (Réserve Naturelle d'Iroise, Bretagne) de l'éradication du rat surmulot (*Rattus norvegicus*). Revue d'Ecologie (Terre Vie) 59: 319–329

Macdonald D, Sidorovich VE, Maran T and Kruuk H (2002) European Mink, *Mustela lutreola*: analyses for conservation. Wildlife Conservation Research Unit, Oxford, UK, 122 pp

Martín A (2002) Rabbit Eradication on Montaña Clara (Canary Islands, Spain). Proceedings of the Workshop on Invasive Alien Species on European Islands and Evolutionary Isolated Ecosystems and Group of Experts on Invasive Alien Species (Horta, Azores). Council of Europe, Strasbourg, tpvs/IAS (2002)2, pp 14–15

Martin JL, Thibault JC and Bretagnolle V (2000) Black rats, island characteristics and colonial nesting birds in the mediterranean: consequences of an ancient introduction. Conservation Biology 14: 1452–1466

Meinesz A (1999) Killer Algae, University of Chicago Press, 376 pp

Oliveira PJ (1999) Habitat restoration on Deserta Grande, Madeira (Portugal): eradication of non-native mammals. In: Proceedings of the Workshop on the Control and Eradication of Non-Native Terrestrial Vertebrates, pp 41–42. CE, Environmental Encounters, No. 41, Council of Europe, Strasbourg, 147 pp

Orueta JF and Ramos YA (1998) Methods to Control and Eradicate Non Native Terrestrial Vertebrate Species. Nature and Environment (Council of Europe) No. 118, Council of Europe, Strasbourg, 66 pp

Pascal M (1999) Eradication of mammals introduced in the islands. In: Proceedings of the Workshop on the Control and Eradication of Non-Native Terrestrial Vertebrates, pp 31–42. CE, Environmental Encounters, No. 41, Council of Europe, Strasbourg, 147 pp

Perfetti A and Sposimo P (2001) Relazione su possibili effetti indesiderati causati dalle operazioni di eradicazione dei ratti (*Rattus rattus*) dell'Isolotto della Scola (LI). Report attached to LIFE-Nature LIFE97NAT/IT/4153 XII.2001

Perfetti A, Sposimo P and Baccetti N (2001) Il controllo dei ratti per la conservazione degli uccelli marini nidificanti nelle isole italiane e mediterranee. Avocetta 25: 126

Pitta Groz M (2002) Invasive alien species as the main threat to Azores seabirds populations. In: Proceedings of the Workshop on Invasive Alien Species on European Islands and Evolutionary Isolated Ecosystems and Group of Experts on Invasive Alien Species (Horta, Azores). Council of Europe, Strasbourg, tpvs/IAS (2002)2, pp 7–8

Rejmanek M and Pitcairn J (2002) When is eradication of exotic pest plants a realistic goal? In: Veitch D and Clout M (eds) Turning the Tide: the Eradication of Invasive Species, pp 249–253. IUCN SSC Invasive Species Specialist Group. IUCN, Gland, Switzerland/Cambridge, UK, viii + 414 pp

Rouland P (1985) Les castors canadiens de la Puisaye. Bulletin Mensuel de l'Office National de la Chasse 91: 35–40

Simberloff D (2002) Today Tiritiri Matangi, tomorrow the world! Are we aiming too low in invasives control? In: Veitch D and Clout M (eds) Turning the Tide: the Eradication of Invasive Species, pp 4–12. IUCN SSC Invasive Species Specialist Group. IUCN, Gland, Switzerland/Cambridge, UK, viii + 414 pp

Smallshire D and Davey JW (1989) Feral Himalayan Porcupines in Devon. Nature in Devon (Journal of the Devon Wildlife Trust) 10: 62–69

Tershy BR, Donlan CJ, Keitt BS, Croll DA, Sanchez JA, Wood B, Hermosillo MA, Howald GR and Biavaschi N (2002) Island conservation in north-west Mexico: a conservation model integrating research, education and exotic mammal eradication. In: Veitch D and Clout M (eds) Turning the Tide: the Eradication of Invasive Species, pp 293–300. IUCN SSC Invasive Species Specialist Group. IUCN, Gland, Switzerland/Cambridge, UK, viii + 414 pp

Wittenberg R and Cock M (2001) Invasive Alien Species: a Toolkit of Best Prevention and Management Practices. GISP/CAB International, Wallingford, UK

Zonfrillo B (2002) Puffins return to Ailsa Craig. Scottish Bird News 66: 1–2

Biological Invasions (2005) 7: 135–140

French attempts to eradicate non-indigenous mammals and their consequences for native biota

Olivier Lorvelec* & Michel Pascal
*INRA, Station SCRIBE, Équipe Gestion des Populations Invasives, Campus de Beaulieu, 35 042 Rennes Cedex, France; *Author for correspondence (e-mail: lorvelec@beaulieu.rennes.inra.fr; fax: +33-2-23485020)*

Received 4 June 2003; accepted in revised form 30 March 2004

Key words: eradication methods, France, impacts, native biota, non-indigenous mammals

Abstract

Many European politicians, managers, and scientists believe that non-indigenous species cannot be eradicated and that attempts to do so are hazardous because of frequent undesirable results. This notion seems to be based on the view that successful eradications undertaken in many other parts of the world cannot be generalised. To allow reasoned consideration of this argument, the eradication of non-indigenous vertebrate species performed in the French territories (European and overseas) and their recorded consequences on native fauna and flora are synthesised. Nineteen vertebrate eradication attempts were recorded, with seven mammal species as the targets. Of these attempts four failed for technical reasons and one for reasons undetermined as yet. These operations took place on islands of four biogeographical areas (West-European, Mediterranean, West Indies and Indian Ocean subantarctic) except a continental one (West-European continent). Among these 19 attempts, 13 were conducted according to a global strategy that provided data on the impact of the disappearance of the non-indigenous species on several native species. This impact, never detrimental, was determined for 14 species (one mammal, nine birds, one marine turtle, one crab, one beetle, one plant). Unexpected consequences of the disappearance of the invader were recorded for four native species (29%). This result highlights the poverty of natural historical information for several taxa and the flimsiness of the empty niche concept that is often used to argue for the delay of or to prevent any action again a non-indigenous species. If French territories can be taken as an example, eradications of non-indigenous species are not impossible; a good risk assessment prevents undesirable long-term consequences for native species and several native species benefited from the disappearance of the invader. Furthermore, eradication constitutes a powerful experimental tool for ecology and natural history studies if conceived as both a management and research operation.

Introduction

In France as in many countries, preventing introductions and managing non-indigenous species are perceived by many people, including scientists, as nearly impossible. Furthermore, when reviews such as the recent one by Simberloff (2002) show that eradicating non-indigenous species is feasible, a response is that foreign examples are of little relevance to France. Another objection to attempts to eradicate non-indigenous species is the claim that, even if such operations are successful, their consequences are hazardous.

Harmful consequences of eradications, even when the project proceeds as expected, have been documented, and we must admit that many eradication projects have not included monitoring the consequences of the disappearance of the target species. However, well-planned eradication projects have been conducted during the last two decades in many nations, but there has been no synthesis of the information provided by such experiments conducted in European countries.

The aim of this paper is to review all attempts to eradicate non-indigenous vertebrates from French territories and to summarise the major expected, unexpected, desirable and undesirable consequences. This synthesis was done with the goal of allowing European politicians, managers and scientists to elaborate a well-informed opinion on the subject, if France may be taken as an example for Europe.

Methods

We reviewed all known eradication attempts performed in France, its overseas departments and territories enclosed.

These operations were classified according to a two step typology. In the first step, a distinction between operations that followed (I) or did not follow (II) a five-point global strategy. This strategy includes a description of the initial situation of several components of the ecosystem (1), the eradication attempt itself (2), monitoring the success or failure of the attempted eradication (3) and, if eradication succeeded, assessing the impact of the eradication on native species (4). The final step in the strategy is the establishment of a control system (5) to prevent new invasions by the target species (Pascal and Chapuis 2000; Courchamp et al. 2003). The second step drew a distinction between attempts that followed (I.a) or did not follow (I.b) an eradication strategy that incorporated two techniques, life-trapping followed by the use of toxic anticoagulants. This strategy reduces by more than 90% the release of toxins into the ecosystem (Pascal et al. 1996).

The only impacts on native species of eradicating non-indigenous species that we considered in this synthesis are those that were quantified. The assessment of these impacts is based on the measurements of four types of variables before and after the target species was removed. The first kind of variable encompasses an abundance index as determined by an exhaustive census (*Pringlea antiscorbutica* in Kerguelen Island) or standardised sampling censuses. The second consists of an exhaustive census of breeding pairs of bird, the third of breeding success for a representative number of nests, and the last of an exhaustive census of destroyed nests.

Expected consequences of an eradication are ones that were predicted, whether or not they are desirable. Unexpected consequences are unpredicted ones. Consequently, several unexpected consequences are not recorded in this study because of a lack of quantification resulting from the absence of pre-eradication data.

Results

Nineteen vertebrate eradication attempts were recorded (Table 1); the target species were all mammals (*Castor canadensis*, *Rattus rattus*, *Rattus norvegicus*, *Mus musculus*, the *Felis silvestris* and *Oryctolagus cuniculus* feral forms, *Herpestes javanicus auropunctatus*).

Among these 19 attempts, 6 did not follow the five-point global strategy. Among the 13 that adhered to this strategy, 10 used life-trapping and toxic baits in succession. One attempt used a virus (myxomatosis), three attempts used exclusively guns and six used exclusively toxic baits.

Only one successful attempt was in the continental area of France. This eradication was undertaken soon (9–10 years) after the Canadian Beaver was first observed in the wild (Rouland 1985). All the other eradication attempts were performed in insular ecosystems.

Seven attempts were devoted to 10 islands with an area of 1 ha or more and eight islets that belong to six oceanic temperate archipelagos on the coast of Brittany. Among these operations is the first one recorded for France (Rouzic Island, Sept-Îles Archipelago) which was performed successfully in a 1951 half-month campaign using an acute poison (strychnine). For all these operations, the target species was the Norwegian rat. In all the cases except for the campaign on Rouzic, the non-indigenous species was eradicated during a single 1-month operation (Pascal et al. 1996; Kerbiriou et al. 2004), and the single recorded failure can be ascribed to failure to follow the eradication protocol.

Three attempts were devoted to eradicating the ship rat from 20 Mediterranean islands or islets on the coast of Corsica. As for the Norwegian rat populations from Brittany islands, the Lavezzu ship rat population was eradicated during

Table 1. Non-indigenous vertebrate eradication attempts in France and its territories.

Continental area				*C. canadensis*	(1)	1984/85	I.b	S	Yes
Oceanic temperate	Archipelago	Island	S. (ha)	Target species	Op.	Year	Typology	Tech.	Success
	Sept-Îles	Rouzic	3.3	*R. norvegicus*	(1)	1951	II	P	Yes
		Bono	22	*R. norvegicus*	(1)	1994	I.a	TP	Yes
		îie aux Moines	9	*R. norvegicus*		1994	I.a	TP	Yes
		île Plate	5	*R. norvegicus*		1994	I.a	TP	Yes
		île aux Rats	0.2	*R. norvegicus*		1994	I.a	TP	Yes
	Rimains	Rimains	1.5	*R. norvegicus*	(1)	1994	I.a	TP	Yes
		Chatellier	1	*R. norvegicus*		1994	I.a	TP	Yes
		Rocher de Cancale	0.2	*R. norvegicus*		1994	I.a	TP	Yes
	Molène	Trielen	17	*R. norvegicus*	(1)	1996	I.a	TP	Yes
		Enez ar C'hrizienn	1.3	*R. norvegicus*		1996	I.a	TP	Yes
	St. Riom	St. Riom +	14.5	*R. norvegicus*	(1)	2000	I.a	TP	No
		6 islets	1.7	*R. norvegicus*		2000	I.a	TP	No
	Houat	île aux Chevaux	2.5	*R. norvegicus*	(1)	2002	I.a	TP	Yes
	Tomé	Tomé	± 30	*R. norvegicus*	(1)	2002	I.a	TP	Yes
Mediterranean	Lavezzi	Lavezzu +	73	*R. rattus*	(1)	2000	I.a	TP	Yes
		16 islets	17	*R. rattus*		2000	I.a	TP	Yes
	Cerbicales	Toro	0.9	*R. rattus*	(1)	1990/91	II	P	Yes
		Folaca	0.2	*R. rattus*	(1)	2001	I.a	TP	Yes?
		Folaccheda	0.1	*R. rattus*	(1)	2001	I.a	TP	No
Tropical	Martinique	Burgaux	0.49	*R. rattus*		1999/01/02	I.a	TP	Yes
		Percé	0.54	*R. rattus*		1999	I.a	TP	Yes
		Hardy	2.63	*R. rattus*		1999/01/02	I.a	TP	Yes
		Poirier	2.1	*R. rattus*		1999/2002	I.a	TP	Yes
	Guadeloupe	Fajou	120	*R. rattus*	(1)	2001/2002	I.a	TP	No
		Fajou	120	*M. musculus*		2001	I.a	TP	Yes?
		Fajou	120	*H. javanicus*		2001	I.a	T	Yes
Sub Antarctic	Kerguelen	Grande Terre	650,000	*O. cuniculus*	(1)	1956	II	V	No
		Grande Terre	650,000	*F. catus*	(1)	1960/71–77	II	S	No
		île Verte	150	*O. cuniculus*	(1)	1992/93	I.b	P	Yes
		île Guillou	140	*O. cuniculus*	(1)	1994/95	I.b	P	Yes
		île Guillou	140	*F. catus*		1994/96	I.b	S	Yes
		île aux Cochons	165	*O. cuniculus*	(1)	1997–2002	I.b	P	Yes
	St. Paul	St. Paul	800	*R. Rattus*	(1)	1996	II	P	Yes
		St. Paul	800	*O. cuiculus*		1996	II	P	Yes

For classification, see the text. Op.: Operation (some took place at the same time in several islands and some had several target species). Tech.: Technique; P: Poison; T: Trapping; S: Shouting; V: Virus.

a one-month operation (Pascal et al., submitted). The only cited failure occurred on an islet lying less than 50 m off Corsica. This islet was probably re-invaded spontaneously by the ship rat after the eradication. Two attempts were devoted to five islands of two tropical archipelagos in the Lesser Antilles. The first was undertaken against the ship rat population of the four islets of the Saint-Anne Archipelago off Martinique. Success was achieved in four yearly 1-month operations (Pascal et al. 2004). The second was devoted to the simultaneous eradication of the small Indian mongoose, the ship rat, and the house mouse from an island off Guadeloupe covered by 110 ha of mangrove and 10 ha of dry forest (Fajou Island). Eradication of the mongoose by trapping alone and perhaps of the house mouse (not totally controlled) by trapping and poison was achieved in a one-month campaign, but the eradication of the ship rat failed after two yearly 1-month attempts, and the reasons for failure have not yet been identified (Lorvelec et al. 2004).

Between the 1950s and the 1970s, several unsuccessful attempts were made to eradicate the feral rabbit and the feral cat from the Mainland of Kerguelen Archipelago in the Indian Ocean subantarctic region using myxomatosis and guns, respectively (Pascal 1983; Chapuis et al. 1994, 1995; Chapuis 1995). Since then, the rabbit and the cat were eradicated by the use of toxic baits and guns, from three and one islands of the Kerguelen Morbihan Gulf, respectively (Chapuis et al. 2001). Also during the 1990s, the ship rat and the rabbit were eradicated from Saint Paul Island by a technique developed in New Zealand: spreading toxic baits by helicopter (Micol and Jouventin 2002).

Only the 13 eradication attempts that adhered to the five-point global strategy provided data that allow assessment of the impact induced by the removal of non-indigenous on several indigenous species. The data in Figure 1 are restricted to the species with populations that reacted significantly to the eradication.

These data concern one mammal (*Crocidura suaveolens*), four terrestrial birds (*Rallus longirostris, Anthus petrosus, Prunella modularis, Troglodytes troglodytes*), five marine birds (*Hydrobates pelagicus, Puffinus lherminieri, Anous stolidus, Sterna anaethetus, Calonectris diomedea*), one marine turtle (*Eretmochelys imbricata*), one crab (*Gecarcinus ruricola*), one beetle (*Canopsis sericea*) and one plant (*Pringlea antiscorbutica*).

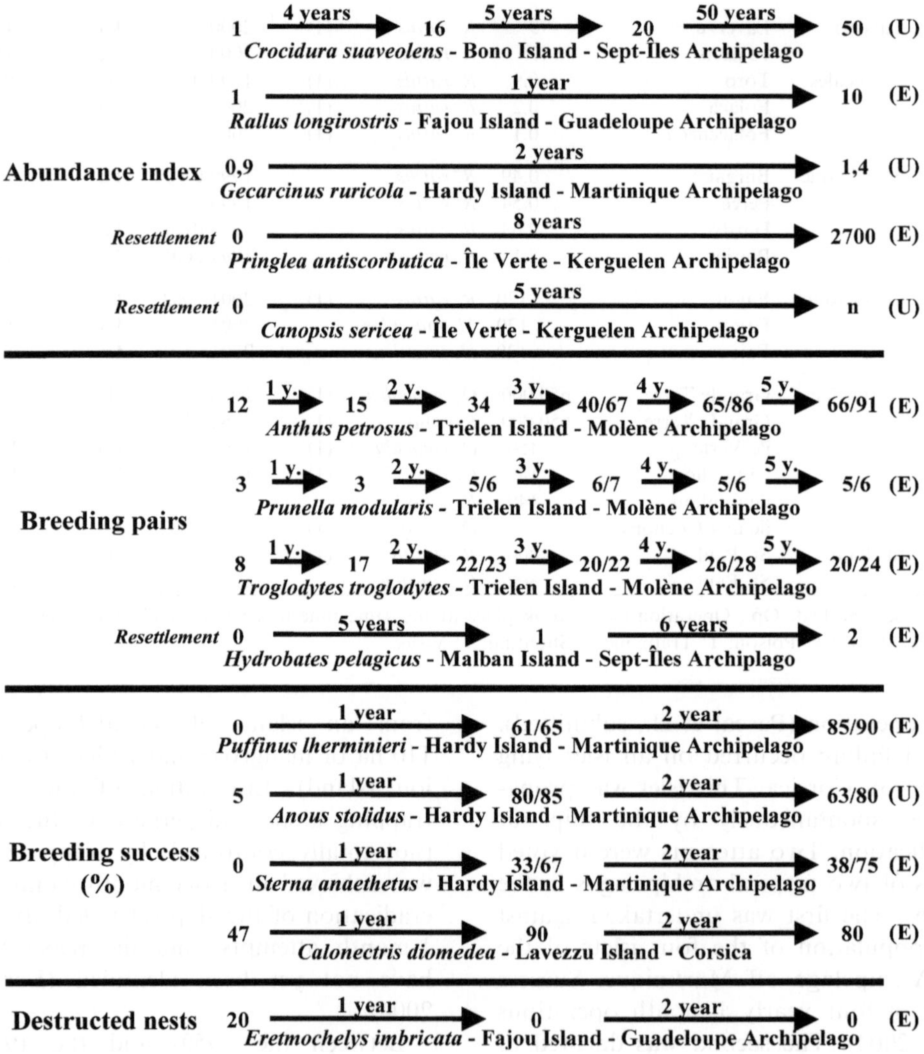

Figure 1. Recorded impacts on native species of eradication of non-indigenous species. (E): expected; (U): unexpected. The quoted time in years corresponds to the period of time between eradication and monitoring.

The impact of the eradication was unexpected for four species among the 14 listed above (29%). Among these four species, the beetle *C. sericea*, which is strictly linked to the Kerguelen cabbage (*P. antiscrobutica*), reinvaded Verte Island in the Kerguelen Archipelago (J.-L. Chapuis, pers. comm.) spontaneously after the predicted re-establishment of the Kerguelen cabbage (Chapuis et al.' 2004). Also unexpected and still unexplained were the high increase of the abundance index (number of capture per trapping night) of the Lesser white-toothed shrew (*C. suaveolens*) after the removal of the Norwegian rat (Pascal et al. 1998) and those of the crab *G. ruricola* after the ship rat eradication (Pascal et al. 2004).

All the evolutions cited in Figure 1 favour the indigenous species, none disfavour them and three indigenous species spontaneously re-established after having disappeared locally.

Some trapping (Pascal et al. 1996; Lorvelec et al. 2004) or poisoning (Pascal et al. 1996; Chapuis et al. 2001) of individuals of both native and non-indigenous non-target species were recorded. Nevertheless, these collateral impacts of eradication were never detectable the year after the eradication.

Discussion

Although the vertebrate target species of the French eradication attempts were all mammals, these operations were performed in five biogeographical areas: West-European continental, West-European islands, Mediterranean islands, West Indies islands and Indian Ocean subantarctic islands. Among the 19 attempts, four failed for technical reasons (the eradication of the feral rabbit and the feral cat from Grande-Terre Island from Kerguelen Archipelago and the eradication of the Norwegian rat from Saint-Riom Archipelago) and one for unknown reasons (the ship rat from Fajou Island). Nevertheless, successes were recorded for each biogeographical region and for several species. These results prove that eradication of non-indigenous species from French ecosystems and, by extension, European ones is not a pipe dream, at least for islands or similarly restricted locations.

None of these attempts produced detectable undesirable long-term consequences for the native biota despite a specific survey devoted to finding such consequences among the 13 eradication operations that followed the five-point global strategy. These results suggest that, at least for the target non-indigenous species, the ecosystems, and the eradication processes discussed above, the risk for native biota is low.

All the recorded consequences of eradicating non-indigenous species favoured native species. Although the majority of these native species are birds (9/14), species from five other taxa benefited from the eradication.

The number of these other taxa recorded as benefiting from eradication would have been larger if the pre-eradication understanding of the interaction between native and non-indigenous species had been greater. Such information would have led to an increase in the number of taxa monitored and, incidentally, to a more precise assessment of the impacts of non-indigenous species. This statement is supported by the unexpected beneficial consequences of the eradication recorded here for four native species, this is one-third of the total species noted. This last result emphasises the flimsiness of the empty niche concept that is often invoked to delay or prevent action against a non-indigenous species, a newcomer or not, before its detrimental effects on the ecosystem functioning are established. Although interactions between pairs of taxa such as rats and marine birds or rabbits and plants are well documented, the natural history of many other interactions is very poorly understood. This situation can be at least partly remedied if each costly eradication operation is conceived as a management and research operation and not as a management operation alone (Pascal et al. 1996; Pascal and Chapuis 2000).

The conclusions of this synthesis devoted to the French eradication attempts mirror those previously published for many other parts of the world (Veitch and Belle 1990; Veitch and Clout 2002): eradication of non-indigenous species is not impossible, a good risk assessment prevents undesirable long-term consequences for native biota (Simberloff 2002), and several native species take advantage of the disappearance of non-indigenous species. Furthermore, eradications constitute a powerful experimental tool for ecology if they are conceived as a research operation as well as a management one.

Acknowledgements

We are grateful to Dan Simberloff who edited the English of this text and suggested several additions that clarified it.

References

Chapuis J-L (1995) Alien mammals in the French Subantarctic Islands. In: Dingwall PR (ed) Progress in Conservation of the Subantarctic Island, Vol 2, pp 127–132. IUCN, Cambridge, UK

Chapuis J-L, Boussés P and Barnaud G (1994) Alien mammals, impact and management in the French Subantarctic islands. Biological Conservation 67: 97–104

Chapuis J-L, Barnaud G, Bioret F, Lebouvier M and Pascal M (1995) L'éradication des espèces introduites, un préalable à la restauration des milieux insulaires. Cas des îles françaises. Nature–Sciences–Sociétés Special Issue: 51–65

Chapuis J-L, Le Roux V, Asseline J, Lefèvre L and Kerleau F (2001) Eradication of the rabbit (*Oryctolagus cuniculus*) by poisoning, on three islands of the subantarctic archipelago of Kerguelen. Wildlife Research 27: 323–331

Chapuis J-L, Frenot Y and Lebouvier M (2004) Recovery of native plant communities after eradication of rabbits from the subantarctic Kerguelen islands, and influence of climate change. Biological Conservation 117: 167–179

Courchamp F, Chapuis J-L and Pascal M (2003) Mammal invaders on islands: impact, control and control impact. Biological Review 78: 347–383

Kerbiriou C, Pascal M, Le Viol I and Garoche J (2004) Conséquences sur l'avifaune terrestre de l'île de Trielen (Réserve Naturelle d'Iroise, Bretagne) de l'éradication du rat surmulot (*Rattus norvegicus*). Revue d'Ecologie (Terre Vie) 59: 319–329

Lorvelec O, Delloue X, Pascal M and Mège S (2004) Impacts des mammifères allochtones sur quelques espèces autochtones de l'Îlet Fajou (Réserve Naturelle du Grand Cul-de-Sac Marin, Guadeloupe), établis à l'issue d'une tentative d'éradication. Revue d'Ecologie (Terre Vie) 59: 293–307

Micol T and Jouventin P (2002) Eradication of rats and rabbits from Saint-Paul Island, French Southern Territories. In: Veitch CR and Clout MN (eds) Turning the Tide: the Eradication of Invasive Species, pp 199–205. IUCN, Cambridge, UK

Pascal M (1983) L'introduction des espèces mammaliennes dans l'Archipel des Kerguelen (Océan Indien Sud). Impact de ces espèces exogènes sur le milieu insulaire. Compte Rendu Société de Biogéographie 59(2): 257–267

Pascal M and Chapuis J-L (2000) Eradication de mammifères introduits en milieux insulaires: questions préalables et mise en application. Revue d'Ecologie (Terre Vie) (Suppl 7): 85–104

Pascal M, Siorat F, Cosson J-F and Burin des Roziers H (1996) Éradication de populations insulaires de Surmulot (Archipel des Sept-Îles – Archipel de Cancale: Bretagne, France). Vie et Milieu – Life and Environment 46(3/4): 267–283

Pascal M, Siorat F and Bernard F (1998) Norway rat and shrews interactions: Brittany. Aliens 8: 7

Pascal M, Brithmer R, Lorvelec O and Vénumière N (2004) Conséquences sur l'avifaune nicheuse de la réserve naturelle des Îlets de Sainte-Anne (Martinique) de la récente invasion du Rat noir (*Rattus rattus*), établies à l'issue d'une tentative d'éradication. Revue d'Ecologie (Terre Vie) 59: 309–318

Rouland P (1985) Les castors canadiens de la Puisaye. Bulletin Mensuel de l'Office National de la Chasse 91: 35–40

Simberloff D (2002) Today Tiritiri Matangi, tomorrow the World! Are we aiming too low in invasives control ? In: Veitch CR and Clout MN (eds) Turning the Tide: the Eradication of Invasive Species, pp 4–12. IUCN, Cambridge, UK

Veitch CR and Belle BD (1990) Eradication of introduced animals from the islands of New Zealand. In: Towns DR, Daugherty CH and Atkinson IAE (eds) Ecological Restoration of New Zealand Islands, pp 137–146. Department of Conservation, Wellington, New Zealand

Veitch CR and Clout MN (eds) (2002) Turning the Tide: the Eradication of Invasive Species. IUCN, Cambridge, UK

Biological Invasions (2005) 7: 141–147

Successful eradication of invasive rodents from a small island through pulsed baiting inside covered stations

Jorge F. Orueta*, Yolanda Aranda, Tomás Gómez, Gerardo G. Tapia & Lino Sanchez-Mármol
*Gestión y Estudio de Espacios Naturales, S.L., c/ Barquillo, 30, 1º D, Hoyo de Manzanares, 28004 Madrid, Spain; *Author for correspondence (e-mail: jorge.orueta@telefonica.es; fax: +34-91-5230897)*

Received 4 June 2003; accepted in revised form 30 March 2004

Key words: anticoagulants, covered baiting stations, Mediterranean, pulsed baiting, *Rattus rattus*, small islands

Abstract

We show the results of an eradication campaign against *Rattus rattus* developed in Rey Francisco Island (12 ha), Chafarinas islands, southwestern Mediterranean. Rat population size was estimated by snap trapping in up to 93.47 ind./ha and a trapping index of 9.58 captures/100 traps-night. We think that population was underestimated because of the number of traps found strung but without capture. Several products were tested in order to define the method of eradication. In 1992, we selected a second generation anticoagulant, pelleted brodifacoum 50 ppm into 5 l plastic containers as baiting stations. Bait consumption reached zero after three pulses, and intensive searching of tracks and signals were unsuccessful. After more than two years of absence of signals and sightings, in 1995, rat scats were observed in Rey Francisco, and the population rose dizzily. After several snap-trapping sessions in 1996, 1997 and 1999, when trapping success reached 37 captures/100 trap-nights, a new campaign started in autumn–winter 1999–2000 using flocoumafen 50 ppm inside 180 baiting stations. Eradication occurred with a very low risk for non-target fauna, setting less than 1 kg/ha of bait each time. Monitoring, both with snap traps and baiting at a lower intensity assures the absence of reinvasion.

Introduction

Introduced predator and herbivore species are one of the main threats affecting threatened bird species in the world, with a much higher proportion of species being at risk on islands (Collar et al. 1994, pp. 24–25); in fact, 90% of extinction cases have occurred on islands (Courchamp et al. 1999, p. 282). Invasive alien mammals have well known effects on previously predator-free isolated ecosystems, being the major cause of extinction and risk (Moors and Atkinson 1984; Atkinson 1985; King 1985; Fritts 1998; Courchamp et al., in press). Seabirds can be seriously affected by alien predators, and they have put some species on the brink of extinction (Collar et al. 1994, pp. 33–37; Menezes and Oliveira 2002; Pierce 2002). Rats are the introduced predators that have reached most islands (Atkinson 1985) and caused most of birds' extinction there (King 1985). They affect many species in most terrestrial vertebrate or invertebrate groups, through depredation and competition for shelter or food (Courchamp et al. 1999, p. 283).

Mediterranean islands have suffered the flow of introductions since pre-Neolithic times with the consequent homogenisation of biodiversity through the extinction of original fauna (Masseti 2002). Throughout the centuries, rats have conditioned the distribution and abundance of seabirds, the effect being more evident on smaller than on bigger islands (Martin et al. 2000).

Several methods have been used to fight commensal rodents on islands, in order to protect native biota (see the review in Orueta and Aranda 2001). The most commonly used method is poisoning with anticoagulants. Second generation anticoagulants were developed to assure a single dose poisoning, because first generation anticoagulants allowed the development of resistance in rats (Greaves and Rennison 1973; Hadler and Shadbolt 1975; Meehan 1984; Greaves et al. 1987). As death takes some days to occur, rodents can continue consuming bait, thus getting an overdose in their tissues and, especially, in their guts which is dangerous to predators and scavengers (Kaukeinen 1982; Merson et al. 1984). To overcome this threat, pulsed baiting is proposed as a method to reduce the amount of poison available in the field, thus minimising secondary hazard. This technique consists in setting the dose that can be consumed in a single event and lets the poison act during several days before subsequent pulses (Dubock 1984).

Chafarinas archipelago (Djafaren) (35°20′ N, 2°25′ W) lies north of Ras el Ma (Cabo de Agua), in the northern coast of Morocco, in the southwestern Mediterranean. Rey Francisco (12 ha) is the easternmost and the smallest of the three islets, very close to Isabel II which is inhabited. It is elongated in shape with four small land masses connected by a narrow isthmus. Vegetation is conditioned by a semi-arid climate, guano and sea influx, with *Lycium intricatum*, *Atriplex halimus* and *Suaeda vera* as dominant species. Several reptiles inhabit the isle, and it has the second most important *Larus audouinii* population in the world.

Chafarinas islands have a whole lot of characteristics that made them suitable for holding rat populations (Atkinson 1985).

To reduce the threat to nesting seabirds and other native fauna, it was decided to use poison with an anticoagulant (Coulter et al. 1985; Moors 1985; Greaves et al. 1987) chiefly a second generation product (Greaves and Rennison 1973; Hadler and Shadbolt 1975; Meehan 1984; Greaves et al. 1987).

In 1992, we carried out a trial to reduce the rat population, and our results were more successful than expected. In order to run the lowest risk for non-target species, we employed pulsed baiting and covered baiting stations. During 1999, a permanent team was present in the archipelago, and there was a longer term commitment to achieve rat eradication.

Materials and methods

Population size estimation

The population size was estimated by snap trapping on four occasions: 1992, 1996, 1997 and 1999. In February 1992, three grids of traps were installed on the island, and traps were layed every 10 m; the networks were formed by $5 * 10$ traps on the northern part of the island, covering 0.36 ha, $5 * 8$ traps in the central section (0.28 ha) and $10 * 10$ in the southern area (0.81 ha). We calculated regression lines of daily captures vs cumulated captures. We then also compared the whole grid vs the 'inner grid' resulting from the elimination of the data of the perimeter traps, to calculate the effect of immigration in the captures. Differences between both sets of data were analysed with the χ^2 test.

The remaining three years, during the last week of September, three lines of 10 trapping stations (with two snap traps each) were used. As in 1992, traps were placed at sunset and reviewed at dawn when any set trap was sprung to avoid undesired effects on birds and reptiles.

All traps were numbered and tied to a wooden peg by a 15 cm string. The bait was unrefined fish oil mixed with sugar and flour to make it stickier, soaked on a piece of lamp wick. Trapping effort was 950 trap-nights in 1992, 300 trap-nights in 1996 and 1999 and 240 trap-nights in 1997.

For all the trapping events, the ratio of captures/100 trap-nights was also calculated. In order to compare with other studies, the index was also transformed following the 'corrected trap-night' index (Houston 2002):

$$\frac{\text{captures}}{\text{trap-nights} - \frac{\text{captures+sprung traps}}{2}} * 100$$

After the poisoning campaign of 1999–2000, trapping has been continued in the same manner and for the same time.

Selection of baits and baiting stations

In 1992, brodifacoum 50 ppm was chosen because of its single dose efficacy (Meehan 1984) and availability. It was accessible in two forms with different appearances: 20 g wax blocks and pelleted fodder in bulk. We also wanted to test two models of baiting stations: (1) wood boxes divided into two longitudinal chambers by a 5 cm high partition: the first chamber with one door on each side and the other where the bait can be set threaded on a tightened wire; (2) 5 l plastic prism shaped containers set horizontally.

During two consecutive nights, we compared consumption of pelleted bait in plastic containers and blocks in a wooden box, both having a similar quantity. During the following two nights, we switched baits, and during the last two nights, both station models were baited with pellets. We recorded whether the bait was totally consumed or not.

In 1999, the bait preferred in 1992 was no longer available. Thus, three second generation anticoagulants were tested: brodifacoum 50 ppm in 20 g wax blocks, bromadiolone impregnated cereal grains in 25 g bags and flocoumafen in 16 g wax–cereal blocks. Ten series of three baiting stations were set: one station in each group containing a dose of one of the baits, changing the order. In a second test, we placed 3 lines of 10 stations, switching the relative situation of each bait.

Poisoning

In 1992, we installed 148 baiting stations on the islet, averaging 25–30 m in distance. Every baiting station was mapped and labelled. One hundred and twenty grams of poison was left inside the station, and after six nights, the remaining bait was measured and removed. Three pulses were done in April, August and October.

In 1999, we installed 180 baiting stations, one for every 25 m. In addition to the plastic containers, we also used plastic boxes with one 50 mm hole on one side. Eleven consecutive pulses were maintained between November 1999 and February 2000.

After this campaign, 66 stations have been permanently set during summer, autumn and winter to monitor achievements.

Results

Population size estimation

Only shot traps with rats or rat remains (depredation ranked 16.7–87.5% of captures) were considered for the analysis.

In 1992, captures were as shown in Table 1. The grid on the central part of the island was not considered to calculate densities. We judged that the differences between the central and the border traps were too big to consider the whole network as representative of the area. Capture rates per 100 trap-nights were 9.58% (1992), 31.33% (1996), 27.5% (1997) and 37% (1999). The highest 'corrected trap-nights' index was 54.01 in 1999.

Between 2000 and spring 2003, no captures were recorded by trapping and neither sightings nor signs of rats detected.

Choice of bait and baiting stations

In 1992, pellets were totally consumed after two nights in whichever station they were installed. Wax blocks were only consumed partially, less than one half in every case. Results from 1999 can be seen in Table 2.

Poisoning

In 1992, the bait load was 17,520 g for the first pulse and 16,800 g for the third one (five stations

Table 1. Analysis of captures taken in 1992 at three different sectors of Rey Francisco Island.

Sector	Captures whole grid	Captures inner grid	χ^2	Regression line	Density, rats/ha
North	34	15	n.s.	$y=-0.8x+26.92$, $r^2=1$	93.47
Middle	14	3	$P<0.001$	$y=-0.89x+11.98$, $r^2=0.99$	–
South	43	22	n.s.	$y=-0.84x+34.77$, $r^2=0.99$	51.09

Table 2. Results of different bait tests done in 1999.

	Day 1		Day 2		Day 3	
	%	W	%	W	%	W
First test						
Brodifacoum wax-block	0	0	75	15	90	18
Bromadiolone cereal	48	12	83	20.75	95	23.75
Flocoumafen wax–cereal block	10	1.7	100	16	100	16
Second test						
Brodifacoum wax-block	3	0.6	13	2.6	93	18.6
Bromadiolone cereal	55	13.75	78	19.5	100	25
Flocoumafen wax–cereal block	93	14.88	100	16	100	16

were lost). The total consumption was 992.5 g/ha. After this campaign, there were no signs of rats during 2 years, until 1995.

As during 1992, bait consumption was much smaller than the loaded bait, we used a smaller quantity of bait per pulse during 1999. To start with, the bait load was 2880 g (one block per station), but after two pulses with consumption close to 100% it was increased to 8592 g (three blocks per station). During this campaign, the total bait consumption was 702.75 g/ha. Detailed results can be seen in Tables 3 and 4. Since the summer of 2000, no consumption of bait has been recorded.

Discussion

Population size estimation

The grid method used in 1992 allowed us to define regressions to calculate densities; they ranked between 93.47 individuals/ha in the northern part of the island and 51.09 in the south. Several facts, such as detected cannibalism on rat corpses and the amount of traps sprung without a catch, indicate that this could be an underestimation, probably due in part to trap-shyness and learning. So, in late winter, the population of black rats can reach very high densities (ca. 100 ind./ha), at least locally.

Trapping during the rest of the years cannot be directly compared with these densities, because we used double traps at each point, and the season was different. Anyway, in 1996, 1997 and 1999, the total amount of captures was much bigger in the same sector of the island that gave 93.47 rats/ha in 1992 (up to 111 rats in 300 trap-nights in 1999 vs 34 rats in 250 trap-nights in 1992). Although we cannot suppose a direct proportion, we can guess that density in the northern part of Rey Francisco at the beginning of autumn could be much greater than 100 rats/ha.

If we compare the capture rate obtained during these last three years with other studies, we see that García et al. (2002) obtained 0.63 rats/trap-hour at Monito island (Puerto Rico); we obtained, approximately, 3 rats/trap-hour in 1999. At Saint-Paul island, Micol and Jouventin (2002) calculated between 5 and 100 rats/ha depending on the location. Hooker and Innes (1995) calculated rat density in a sector of North Island, New Zealand; Shanker (2000) got up to 44 rats/ha in continental locations in India.

The corrected trap-nights index is smaller than figures calculated on a small (5.2 ha) flat island in Fiji for *Rattus exulans* being 65.93 (Houston 2002) to 114.3 (Watling 2002).

Table 3. Results of the campaign done in 1992.

1992	April	August	October
Total load (g)	17,520	17,160	16,800
Consumption (g)	5550	5850	510

Table 4. Results of the campaign done in 1999 and 2000.

1999–2000	1	2	3	4	5	6	7	8	9	10	11
Total load (g)	2880	2880	2880	8592	8624	5216	8160	8640	8640	8218	8256
Consumption (g)	2880	2848	1584	928	144	16	0	0	0	16	32

Bait and baiting stations

As one of the targets of the campaigns was to leave as less poison as possible available in the rats' guts, we considered positive the fact that flocoumafen wax-cereal blocks were consumed faster than the others. This is due to both its palatability (at least as palatable as the cereal alone and very much more than wax blocks) and to the smaller dose. Cereals could be more easily taken by ants and introduce the product into the non-target food chain.

Although our baiting stations were costlier than others such as pipes or plastic bottles (Garcia et al. 2002; Merton et al. 2002), it was considered that they were safe enough to avoid all non-target risks.

Poisoning

In 1992, close to 992 g/ha of brodifacoum 50 ppm was consumed. The fact that during two years there were no signs of rats and, after first tracks in 1995 a demographic explosion occurred, allows us to think that eradication was probably successful but re-colonisation happened. In fact, the inhabited Isabel II island is less than 200 m close to Rey Francisco. At the time of this first campaign, a dump was very close to the closest point between the two islands. By 1994, the dump was translated to the other side of Isabel II, and incineration became general for organic items. Unfortunately, we were not able to monitor the period after this campaign, and recovery of the rat population, either by remnants or re-colonisation, occurred.

It is notable that in the northern part of the island, where the trapping grid occupied most of the surface in 1992, consumption was quite smaller. Snap-trapping can notably reduce the rat population if the area is small. This was already noted by Moors (1985) in an islet where baiting was unnecessary after trapping.

In the 1999–2000 campaign, about 700 g/ha of flocoumafen 50 ppm was consumed in total, while the maximum amount of poison available at each moment was about the same quantity. Merton et al. (2002) used approximately 600 g/ha of brodifacoum 50 ppm at a first pulse, killing apparently most of the rats and approximately 2 kg/ha in total. In open air poisoning campaigns, the available bait at any moment is higher than that inside baiting stations. The doses used on Rey Francisco Island campaigns ranged 35–50 mg of anticoagulant per hectare, and García et al. (2002) used 24.5 mg/ha of brodifacoum. Normal rates for anticoagulants in aerial application exceed 200 mg/ha. For example, in Saint-Paul (Micol and Jouventin 2002), the rate ranked from 10 to 40 kg/ha; bromadiolone 20 ppm was spread at 10 kg/ha on Browns Island (Veitch 2002a) and on Fanal Island, the dose was 10 kg/ha of brodifacoum 20 ppm (Veitch 2002b).

The use of covered baiting stations and pulsed baiting notably reduced the amount of bait exposed in the field, although it is more work costly. In Chafarinas, it was cost effective because at least two people have to be present anyway on the archipelago permanently. This task could be accomplished easily among their other duties during autumn and winter, when there are less activities to be done. It can be used for small islands, although probably it could be useful at a bigger scale if personnel are available. In Congreso, one of the islands of the archipelago, despite a similar effort, we did not arrive at a solution to eradicate the population; this was mainly because of the inaccessible cliffs, although the effect on seabirds was highly positive (Orueta et al. 2002) and that several indexes (snap traps, bait consumption) proved that the population was very close to being exterminated. It is worth highlighting the lack of corpses although research was done to recover any possible dead rats. No secondary data on non-target deaths were registered in any of the campaigns. It is possible that the disappearance of some baits during the last two pulses in 2000 after three pulses with no consumption at all (Table 4) could be due to *Larus michahellis*, but it is not sure that consumption occurred. Anyway, we did not find any corpse of this very common species during the campaigns and monitoring.

Monitoring of trapping has been conducted since 2000, and no rats have been recorded. There are also 60 baiting stations homogeneously distributed on the island, and no bait consumption was registered up to spring 2003.

In 1999, the greater density of baiting stations and the proximity of pulses (without letting the population to recover) were factors contributing to success. The continuity of baiting throughout the years is a guarantee against re-invasion, but an intensive poisoning campaign should be done in Isabel II, as well as quarantine measures should be taken to minimise risk in the future.

Acknowledgements

We acknowledge all the persons who have contributed in the field campaigns during all these years, as well as the anonymous reviewer who made valuable remarks. We also want to recognise the role played by the military personnel on Chafarinas islands. Georgina Alvarez, Jorge Moreno and Javier Zapata (Ministry of Environment) were the persons in charge of the Chafarinas protected area. This work has been done thanks to the economical and logistic support of the Organismo Autónomo de Parques Nacionales.

References

Ahmed MS and Fiedler LA (2002) A comparison of four rodent control methods in Philippine experimental rice fields. International Biodeterioration & Biodegradation 49: 125–132

Atkinson IAE (1985) The spread of commensal species of Rattus to oceanic islands and their effects on island avifaunas. In: Moors PJ (ed) Conservation of Island Birds. ICBP Technical Publication 3, pp 35–81. ICPB, Cambridge, UK

Collar NJ, Crosby MJ and Stattersfield AJ (1994) Birds to Watch 2. The World List of Threatened Birds. BirdLife International, Cambridge, UK

Coulter MC, Cruz F and Cruz J (1985) A programme to save the dark-rumped petrel, Pterodroma phaeppygia, on Floreana Island, Galapagos, Ecuador. In: Moors PJ (ed) Conservation of Island Birds. ICBP Technical Publication 3, pp 177–180. ICPB Cambridge, UK

Courchamp F, Chapuis JL and Pascal M (In press) Mammals invaders on islands: impact, control and control impact. Biological Reviews

Courchamp F, Langlais M and Sugihara G (1999) Cats protecting birds: modelling the mesopredator release effect. Journal of Animal Ecology 68: 282–292

Dubock AC (1984) Pulsed baiting: a new technique for high potency slow acting rodenticides. Proceedings of the Conference on Organisation and Practice of Vertebrate Pest Control 10: 123–136

Fritts TH (1998) The role of introduced species in the degradation of island ecosystems: a case study of Guam. Annual Review of Ecology and Systematics 9: 113–140

García MA, Díez CE and Alvarez AO (2002) The eradication of *Rattus rattus* from Monito Island, West Indies. In: Veitch CR and Clout MN (eds) Turning the Tide: the Eradication of Invasive Species, pp 116–119. IUCN SSC Invasive Species Specialist Group. IUCN, Gland, Switzerland/Cambridge, UK

Greaves JH and Rennison BD (1973) Populations aspects of warfarin resistance in the brown rat, *Rattus norvegicus*. Mammal Review 3: 27–39

Greaves JH, Redfern R, Ayres PB and Gill JE (1987) Warfarin resistance: a balanced polymorphism in the Norway rat. Genetical Research 30: 257–263

Hadler MR and Shadbolt RS (1975) Novel 4-hydroxycoumarin anticoagulants active against resistant rats. Nature 253: 275–277

Hooker S and Innes J (1995) Ranging behaviour of forest-dwelling ship rats, *Rattus rattus*, and effects of poisoning with brodifacoum. New Zealand Journal of Zoology 22: 291–304

Houston DM (2002) Eradicating rats from Maninita Island, Vava'u, Kingdom of Tonga August 2002. New Zealand Agency for International Development, Tonga Visitors Bureau, Ministry of Land, Survey and Natural Resources, Department of Environment, Kingdom of Tonga

Kaukeinen DE (1982) A review of the secondary poisoning hazard to wildlife from the use of anticoagulant rodenticides. Proceeding of a Vertebrate. Pest Conference 10: 151–158

King WB (1985). Island birds: will the future repeat the past? In: Moors PJ (ed) Conservation of Island Birds. ICBP Technical Publication 3, pp 3–15. ICPB, Cambridge, UK

Martin JL, Thibault JC and Bretagnolle V (2000) Black rats, island characteristics and colonial nesting birds in the Mediterranean: consequences of an ancient introduction. Conservation Biology 14(5): 1452–1466

Masseti M (2002) The non-flying terrestrial mammals of the Mediterranean islands: an example of the role of the biological invasion of alien species in the homogenisation of biodiversity. In: Proceedings of the Workshop on Invasive Alien Species on European Islands and Evolutionary Isolated Ecosystems, Horta Açores, 10–12 October 2002, pp 13–15. Council of Europe, Strasbourg, France

Meehan AP (1984) Rats and Mice. Their Biology and Control. Rentokil Ltd, East Grinstead, UK, 383 pp

Menezes D and Oliveira P (2002) Control of introduced predators and herbivores to protect critical species: the case study of the Freira da Madeira. In: Proceedings of the Workshop on Invasive Alien Species on European Islands and Evolutionary Isolated Ecosystems, Horta Açores, 10–12 October 2002, pp 18–19. Council of Europe, Strasbourg, France

Merson MH, Ryers RE and Kaukeinen DE (1984) Residues of the rodenticide brodifacoum in voles and raptors after orchard treatment. Journal of Wildlife Management 48(1): 212–216

Merton D, Climo G, Laboudallon V, Robert S and Mander C (2002) Alien mammal eradication and quarantine on inhabited islands in the Seychelles. In: Veitch CR and Clout MN (eds) Turning the Tide: the Eradication of Invasive Species, pp 182–198. IUCN SSC Invasive Species Specialist Group. IUCN, Gland, Switzerland/Cambridge, UK

Micol T and Jouventin P (2002) Eradication of rats and rabbits from Saint-Paul Island, French Southern Territories. In: Veitch CR and Clout MN (eds) Turning the Tide: the Eradication of Invasive Species, pp 199–205. IUCN SSC Invasive Species Specialist Group. IUCN, Gland, Switzerland/Cambridge, UK

Moors PJ (1985) Eradication campaigns against *Rattus norvegicus* on the Noises Islands, New Zealand, using brodifacoum and 1080. In: Moors PJ (ed) Conservation of Island Birds. ICBP Technical Publication 3, pp 143–155. ICBP, Cambridge, UK

Moors PJ and Atkinson AE (1984) Predation on seabirds by introduced animals, and factors affecting its severity. ICBP Technical publication 2, pp 667–690. ICPB, Cambridge, UK

Orueta JF and Aranda Y (2001) Methods to control and eradicate non native terrestrial vertebrates species. Convention on the Conservation of European Wildlife and Natural Habitats (Bern Convention), Council of Europe Publishing. Nature and Environment Series 118. Strasbourg, France

Orueta JF, Igual M, Gómez T, Tapia GG and Mármol LS (2002) Rat predation on seabirds and control measures in Chafarinas Islands. Workshop on Invasive Alien Species on European Islands and Evolutionary Isolated Ecosystems. Horta, Açores, 10–12 October 2002, pp 17–18. Document T-PVS/IAS (2002) Council of Europe, Strasbourg, France

Pierce RJ (2002) Pacific rats: their impacts on two small seabird species in the Hen and Chickens Islands, New Zealand. In: Veitch CR and Clout MN (eds) Turning the Tide: the Eradication of Invasive Species, p 411. IUCN SSC Invasive Species Specialist Group. IUCN, Gland, Switzerland/Cambridge, UK

Shanker (2000) Small mammal trapping in tropical montane forests of the Upper Nilgiris, southern India: an evaluation of capture-recapture models in estimating abundance. Journal of Biosciences 25(1): 99–111

Veitch CR (2002a) Eradication of Norway rats (*Rattus norvegicus*) and house mouse (*Mus musculus*) from Browns Island (Motukorea), Hauraki Gulf, New Zealand. In: Veitch CR and Clout MN (eds) Turning the Tide: the Eradication of Invasive Species, pp 350–352. IUCN SSC Invasive Species Specialist Group. IUCN, Gland, Switzerland/Cambridge, UK

Veitch CR (2002b) Eradication of Pacific rats (Rattus exulans) from Fanl Island, New Zealand. In: Veitch CR and Clout MN (eds) Turning the Tide: the Eradication of Invasive Species, pp 357–359. IUCN SSC Invasive Species Specialist Group. IUCN, Gland, Switzerland/Cambridge, UK

Watling D (2002) Baseline Survey of Maninita Island, Vava'u, Kingdom of Tonga. Environment Consultants, Fiji

MIX
Papier aus verantwortungsvollen Quellen
Paper from responsible sources
FSC® C105338

If you have any concerns about our products,
you can contact us on
ProductSafety@springernature.com

In case Publisher is established outside the EU,
the EU authorized representative is:
Springer Nature Customer Service Center GmbH
Europaplatz 3, 69115 Heidelberg, Germany

Printed by Libri Plureos GmbH
in Hamburg, Germany